MICROTECHNOLOGY AND MEMS

Springer
Berlin
Heidelberg
New York
Hong Kong
London
Milan
Paris
Tokyo

Physics and Astronomy ONLINE LIBRARY

springeronline.com

MICROTECHNOLOGY AND MEMS

Series Editor: H. Baltes D. Liepmann

The series Microtechnology and MEMS comprises text books, monographs, and state-of-the-art reports in the very active field of microsystems and microtechnology. Written by leading physicists and engineers, the books describe the basic science, device design, and applications. They will appeal to researchers, engineers, and advanced students.

Mechanical Microsensors
By M. Elwenspoek and R. Wiegerink

CMOS Cantilever Sensor Systems
Atomic Force Microscopy and Gas Sensing Applications
By D. Lange, O. Brand, and H. Baltes

Micromachines as Tools for Nanotechnology
Editor: H. Fujita

Modelling of Microfabrication Systems
By R. Nassar and W. Dai

Laser Diode Microsystems
By H. Zappe

Silicon Microchannel Heat Sinks
Theories and Phenomena
By L. Zhang, K.E. Goodson, and T.W. Kenny

L. Zhang
K.E. Goodson
T.W. Kenny

Silicon Microchannel Heat Sinks

Theories and Phenomena

With 106 Figures

Dr. Lian Zhang
Molecular Nanosystems
977 Commercial Street
Palo Alto, CA 94303
USA
Email: lian@stanfordalumni.org

Professor Thomas W. Kenny
Stanford University
Department of Mechanical Engineering
Terman 540
Stanford, CA 94305
USA
Email: kenny@mems.stanford.edu

Professor Kenneth E. Goodson
Stanford University
Department of Mechanical Engineering
Building 500, Room 500S
Stanford, CA 94305
USA
Email: goodson@stanford.edu

Series Editors:

Professor Dr. H. Baltes
ETH Zürich, Physical Electronics Laboratory
ETH Hoenggerberg, HPT-H6, 8093 Zürich, Switzerland

Professor Dr. Hiroyuki Fujita
University of Tokyo, Institute of Industrial Science
4-6-1 Komaba, Meguro-ku, Tokyo 153-8505, Japan

Professor Dr. Dorian Liepmann
University of California, Department of Bioengineering
466 Evans Hall, #1762, Berkeley, CA 94720-1762, USA

ISSN 1439-6599

ISBN 3-540-40181-4 Springer-Verlag Berlin Heidelberg New York

Cataloging-in-Publication Data applied for. Bibliographic information published by Die Deutsche Bibliothek
Die Deutsche Bibliothek lists this publication in the Deutsche Nationalbibliografie; detailed bibliographic data is available in the Internet at http://dnb.ddb.de.

Springer-Verlag is a part of Springer Science+Business Media

springeronline.com

Printed in Germany

Typesetting by the authors
Cover concept: eStudio Calamar Steinen
Cover production: *design & production* GmbH, Heidelberg

Printed on acid-free paper 57/3141/tr - 5 4 3 2 1 0

Preface

There is significant current interest in new technologies for IC (Integrated Circuit) cooling, driven by the rapid increase in power densities in ICs and the trend towards high-density electronic packaging for applications throughout civilian and military markets. In accordance with Moore's Law, the number of transistors on Intel Pentium microprocessors has increased from 7.5×10^6 in 1997 (Pentium II) to 55×10^6 in 2002 (Pentium 4). Considering the rapid increase in the integration density, thermal management must be well designed to ensure proper functionality of these high-speed, high-power chips. Forced air convection has been traditionally used to remove the heat through a finned heat sink and fan module. Currently, with 82 W power dissipation rate, approximately 62 W/cm^2 heat flux, from a Pentium 4 processor with 3.06 GHz core frequency, the noise generated from high rotating speed fans is approaching the limit of acceptable level for humans. However, the power dissipation from a single cost-performance chip is expected to exceed 100 W/cm^2 by the year 2005, when the air cooling has to be replaced by new cooling technologies. Among alternative cooling methods, the two-phase microchannel heat sink is one of the most promising solutions. Understanding the boiling process and the two-phase flow behavior in microchannels is the key to successful implementation of such a device.

This book is based on the research conducted in Micro Structures and Sensors Laboratory, Microscale Heat Transfer Laboratory, and Microfluidics Laboratory at Stanford University, where a closed-loop silicon microchannel two-phase cooling system was invented. Although microchannel heat sink has received more and more attention during the past two decades, most of existing research focused on single-phase flow and modeling, and experimental study is especially insufficient in the two-phase flow and heat transfer area. This book focuses on the phase change phenomena and the heat transfer in sub-150 μm diameter microchannels, with emphases on experimental system design, heat transfer measurement and modeling, and the impact of small dimensions on boiling regimes as well as bubble nucleation mechanisms.

The detailed discussion of experimental system design and presentation of experimental results in this book will benefit thermal scientists and engineers who have research interests in microscale heat transfer. MEMS (Micro-Electro-Mechanical Systems) technology can significantly facilitate the study in heat transfer area by providing instrumented platforms with multiple control and sensing capabilities. Scientists, engineers, and graduate students in thermal and MEMS areas will find the discussion in the book inspiring for their research in microchannel flows and microscale heat transfer in general. This book also includes large amounts of experimental data, as a supplement to existing heat transfer textbooks and references.

It has been a great pleasure to work with a group of talented students and scholars at Stanford University. We would especially like to thank Dr. Jae-Mo Koo and Dr. Linan Jiang for their contributions in the heat transfer modeling; Evelyn Wang and Shilajeet Benerjee for their assistance in the microchannel flow experiments; Prof. Juan Santiago, Dr. James Mikkelsen and Dr. Shulin Zeng for their discussions on electroosmotic pumps. James Maveety and Eduardo Sanchez at Intel Corporation provided invaluable advice on high-power chip thermal management. We are also grateful to James McVittie, Peter Griffin, Robin King, Nancy Latta, and Mahnaz Mansourpour for their constructive suggestions and help in the device fabrication processes. Finally, we would like to thank DARPA HERETIC Program, Intel Corporation, and Stanford Graduate Fellowships for the project funding and support. The device fabrication was accomplished at Stanford Nanofabrication Facilities, funded by the National Science Foundation.

Department of Mechanical Engineering
Stanford University
September 2003

Lian Zhang
Kenneth E. Goodson
Thomas W. Kenny

Contents

1 Introduction

1.1 Overview: Moore's Law and IC Cooling

In 1965, Dr. Moore of Intel Corporation predicted that the number of transistors on an integrated circuit would double every 18 months, known as "Moore's Law" [1.1]. Since then, through continual improvements of the manufacturing technologies, Moore's Law has been successfully maintained until today. As an example in Fig. 1.1, the number of transistors on an Intel CPU (Central Processing Unit) chip increased from 2,250 in 1971 (4004) to 3.1×10^6 in 1993 (Pentium); and now there are 55×10^6 transistors on a latest Pentium 4 processor (3.06 GHz core frequency), introduced in November 2002 [1.2].

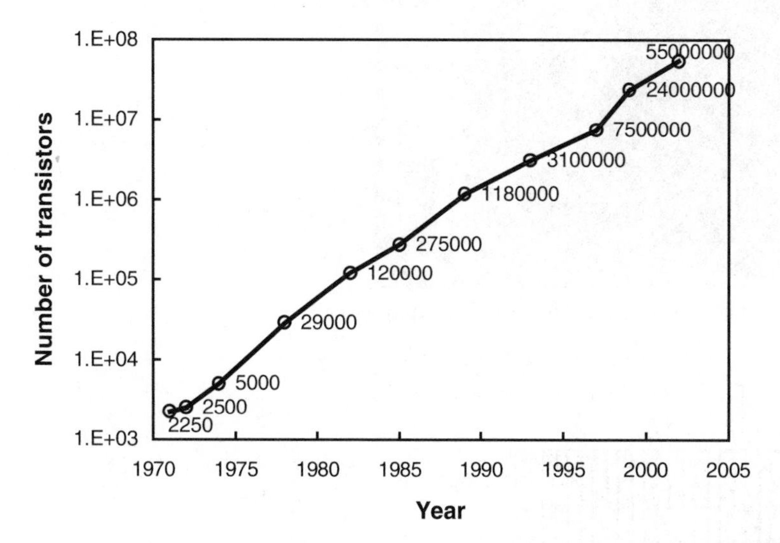

Fig. 1.1 The number of transistors integrated on a single Intel CPU chip reflects Moore's Law [1.1].

The rapid increase in the number of on-chip transistors challenges the packaging technology, especially thermal management, by allowing more and more power into a single chip. Therefore, IC (Integrated Circuit) chip cooling has received increasing attention during the past decade. Forced air convection through heat sinks and fans has been introduced for CPU cooling since early

1990s. Figure 1.2 (a) is the image of a Pentium 4 processor with an attached heat sink and a cooling fan module. As shown in Fig. 1.2 (b), the chip is bonded to a flip-chip carrier and thermal epoxy is applied to the back side to attach a matched material lid as the base of the heat sink. A CPU fan is used to deliver forced air flow to the heat sink. Patel has given an extensive overview on the history of forced air convection for CPU chips, including the components and a discussion of the thermal resistances as shown in Fig. 1.2 (c) [1.3]. In the thermal circuit, $R_{interface}$ and $R_{heatsink}$ represent the conduction in the packaging interface between the chip and the heat sink and the convection through the heat sink. The maximum allowable chip temperature and the ambient air temperature are specified as 85 °C and 40 °C respectively by the manufacturer. In order to satisfy the temperature requirements, the total allowable thermal resistance is limited by $R_{interface} + R_{heatsink} \leq (T_{chip} - T_{ambient}) / power$, that is, $R_{interface} + R_{heatsink} \leq 45$ (℃) / power (W).

(a) The fan and heat sink attached to a Pentium 4 processor.

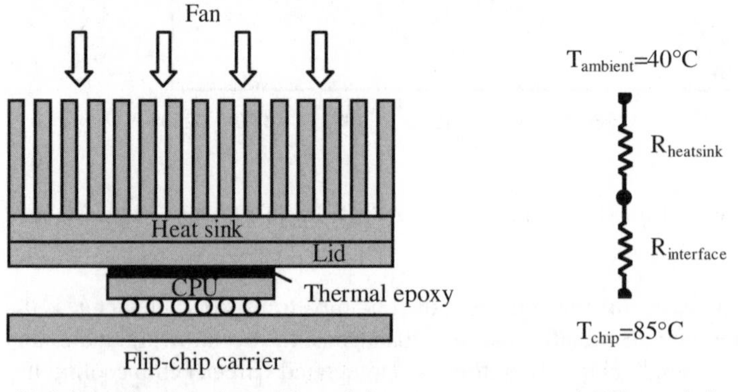

(b) Schematic of the heat sink module.

(c) Thermal circuit.

Fig. 1.2 Forced convective air cooling for a CPU chip and the thermal circuit.

In early 1990s, because of the low chip power dissipation (<15 W), the pin-grid-array packaging with $R_{interface}$≈1.5 °C/W only requires $R_{heatsink}$ to be around 1-1.5 °C/W. Since the mid 1990s, flip-chip packaging has significantly reduced the interface resistance because now the back side of the CPU chip is available as a part of the heat transfer path. With a typical 40×40×4 mm matched material lid (210 W/m-K thermal conductivity), 0.05 mm thick thermal epoxy, and 18×18×0.5 mm chip size, the interface resistance $R_{interface}$ is only approximately 0.25 °C/W. But the chip power has been rapidly increasing during the years. The current CPU chips typically have 50-80 W heat dissipation rate, giving a total allowable thermal resistance of 0.56 °C/W; that is, $R_{heatsink}$ is required to be smaller than 0.3 °C/W. Today, the manufacturer recommended finned heat sink for the state-of-the-art Pentium 4 processor in the 478-pin package with 3.06 GHz core frequency is on the order of 70×70×30 mm in size. The CPU fan is around 70×70×30 mm, providing approximately 10^6 ml/min air flow rate at the rotating speed of 3500-6000 rpm. Since the 35-45 dB fan noise is significant to human hearing, the fans are operated in a variable speed mode and the highest speed is only reached at a certain temperature set point of the package, defined by the manufacturer.

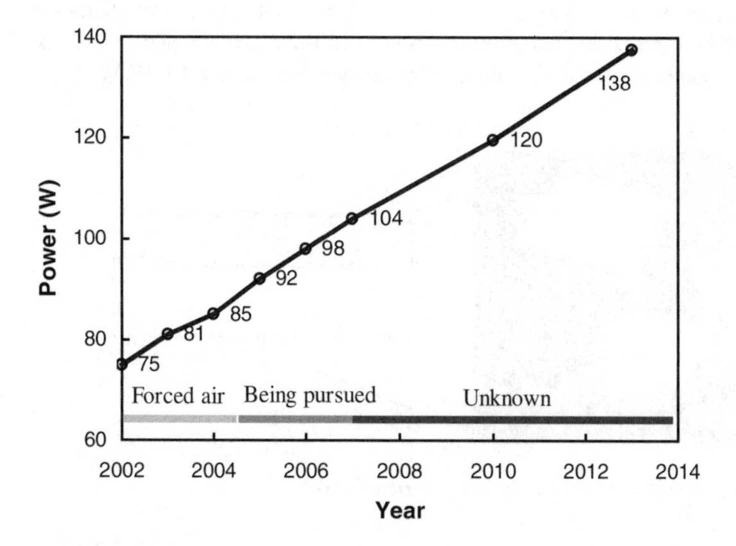

Fig. 1.3 IRTS predictions for packaging and assembly requirements for single cost-performance IC chips, 2002 update[1.4].

However, along with further shrinkage of the CPU core size and the integration of on-chip cache, the chip becomes a nonuniform heat source, which could result in the increase of $R_{interface}$ up to 0.35 °C/W. With the expected 100 W/cm^2 core power dissipation in the next few years, less than 0.1 °C/W heat sink resistance will be required. In other words, a large 150×150×30 mm heat sink will be expected soon. Yet this is not the end of the story—according to the ITRS (International Technology Roadmap for Semiconductors) [1.4], the amount of power dissipation from a single cost-performance chip is increasing linearly with the time and the existing packaging methods are no longer expected to work for new chips starting the year 2007.

As shown in Fig. 1.3, the current heat dissipation rate of 82 W (an Intel Pentium 4 process with 3.06 GHz core frequency) is approaching the limit of forced air convection. Due to the poor thermal transport properties of air, higher heat rates require larger air flow rates, hence larger fans, but the noise and vibration generated by fans will eventually exceed the limit allowed by human hearing. Some improvements for the current existing technologies are under development, such as optimized designs of finned metal heat sinks and thermal considerations in the IC layout design. However, forced air convection is predicted to hardly achieve more than 100 W/cm^2 heat removal. Therefore, in the next few years, the chip power dissipation level will call for new thermal management technologies.

Possible methods include thermoelectric cooling, direct immersion pool boiling, heat pipes, liquid jet impingement or spray cooling, and forced convective boiling in microchannels. Morgan et al. [1.5], Zhou et al. [1.6], and Mudawar [1.7-1.8] compared the various cooling techniques in their review papers. All the techniques can potentially achieve a thermal resistance well below 0.1 °C/W.

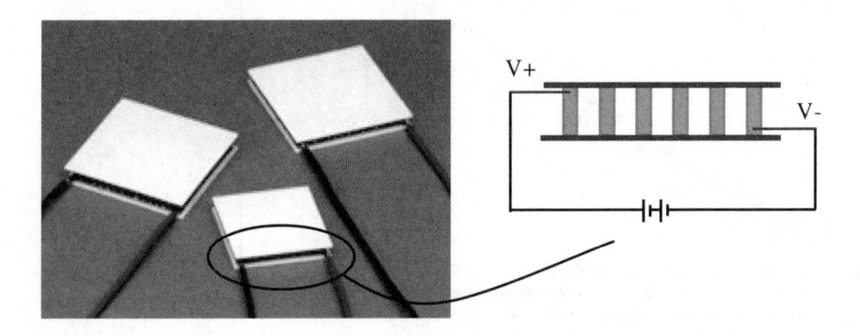

Fig. 1.4 The image and schematic of thermoelectric coolers.

Based on the Peltier effect, thermoelectric coolers, as shown in Fig. 1.4, consist of PN junctions and a DC power supply. As electrons are pumped from one type of semiconductor to another, energy is absorbed from the cold junction and released at the hot junction. Hence the junctions can be used in either cooling or heating applications. The coolers are commercially available. The advantages of this device include that it is very compact, usually less than 60×60×5 mm in size;

there are no moving parts; and it can operate at temperatures below the ambient temperature at the cold junction, which helps to improve device lifetime in some applications such as optical telecommunications. However, the power consumption of these coolers is excessive—a Melcor™ UT8-12-40-F1 thermoelectric cooler that removes 78 W of heat, requires 128.8 W (8 A, 16.1 V) of electrical energy. In fact it is typical that these coolers consume more amount of power to absorb the heat dissipated from the IC chip; therefore, more than twice the amount of heat power has to be removed from the heat rejection side. Due to this limit, thermoelectric coolers are not expected to function well with more than 100 W power chips.

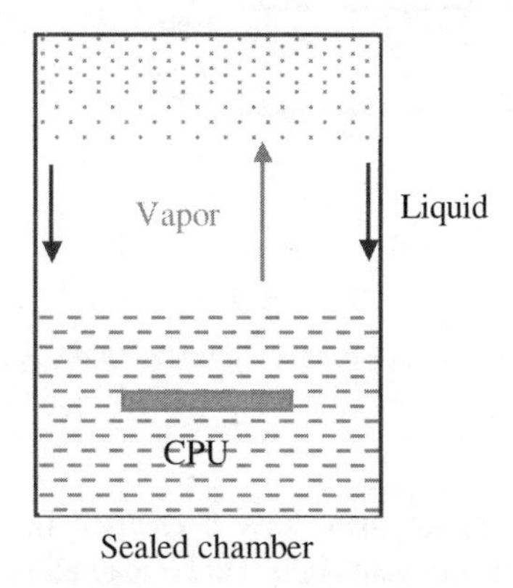

Fig. 1.5 Schematic of a pool boiling system.

Pool boiling relies on the passive circulation of a dielectric coolant inside a sealed chamber, which has been proposed for multi-chip electronic module cooling [1.9-1.11]. As the schematic in Fig. 1.5 shows, the IC chip is completely immersed in a liquid pool. The vapor moves up due to the buoyancy force; and upon condensation, the liquid falls back to the pool under gravity. Pool boiling can achieve large heat transfer rates from liquid-vapor phase change and keep the chip temperature uniform at a nearly constant temperature. With dielectric coolant, cooling of 20 W/cm^2 heat flux has been accomplished; and up to 60 W/cm^2 is expected with liquid subcooling (the liquid pool is maintained at a temperature below the boiling point of the liquid) [1.9]. But the immersion of IC chips in a dielectric liquid and the poorly understood phase change mechanisms of dielectric coolants make the system difficult to implement.

Heat pipes have been commercially used in notebook thermal management applications [1.12-1.16]. Figure 1.6 shows an imbedded heat pipe system in a Machintosh Powerbook with a 400 MHz G3 processor. As shown in the schematic, the heat pipe consists of a vapor channel surrounded by a wick structure. The working fluid circulates between the evaporating region and the condensing region driven by the capillary force, and the heat is absorbed through liquid-vapor phase change. The heat pipes currently used in laptop computers remove about 10 W/cm^2 heat flux. Optimized capillary pumped loops, an extension of heat pipe technology, are expected to remove even higher heat fluxes [1.17-1.18]. However, the volume of the working fluid that can be contained in the wick structure is very limited, which limits the heat pipes from being used in high power applications.

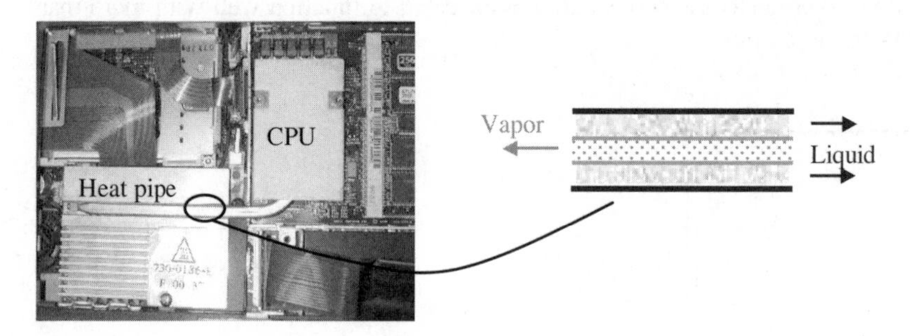

Fig. 1.6 A heat pipe system imbedded in a Machintosh Powerbook computer with a 400 MHz G3 processor.

Forced convection in channels and liquid jet impingement have been used in large scale cooling in the industry for decades. Twenty years ago, Tuckerman and Pease first proposed the idea of silicon microchannel heat sinks, and predicted that single-phase forced convective cooling in microchannels should be feasible for circuit power densities of more than 1000 W/cm^2 [1.19]. Three years later, Kiper predicted more than 500 W/cm^2 heat flux removal with microscale direct liquid impingement to IC chips from an orifice plate [1.20]. When combined with boiling, Ma and Bergles demonstrated 70 W/cm^2 heat flux removal with FC-72 [1.21]. In addition, spray cooling has been proven to achieve even higher heat removal capabilities [1.22-1.24], with up to 1000 W/cm^2 heat flux [1.24], in the two-phase regime.

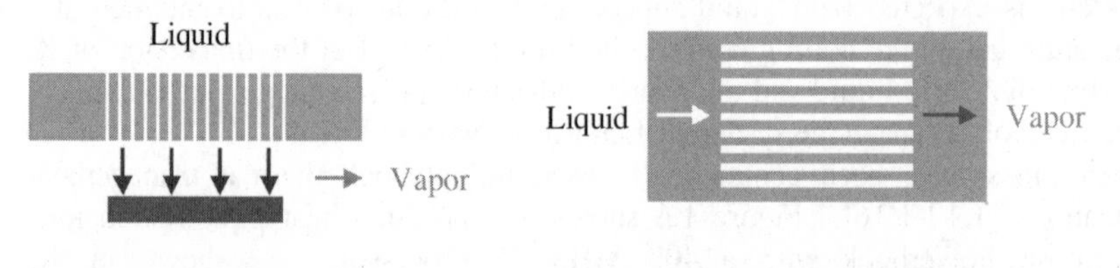

(a) Micro jet impingement. (b) Microchannel heat sink.

Fig. 1.7 Silicon micro jet and microchannel heat sinks.

With today's MEMS (Micro-Electro-Mechanical Systems) technology, it is possible to make micromachined heat sinks such as micro jets and microchannels shown in Fig. 1.7. Because of the material and process compatibility, these micro heat sinks can locally deliver liquid to the IC chip and remove the heat from the chip; even the integration of the heat sink with the IC chip is possible. Macroscale liquid jet impingement has been designed for multi-chip modules [1.25]. Micromachined air jets on the order of 150-500 μm have been demonstrated by several research groups [1.26-1.28]. The most recent experiments show that two-phase liquid jet impingement is expected to remove more than 100 W/cm^2 heat flux at water flow rates below 15 ml/min [1.29-1.30]. In Tuckerman and Pease's experiments, a microchannel heat sink has been demonstrated to remove 790 W/cm^2 with 71 °C temperature increase at 600 ml/min liquid flow rate [1.19]. Mudawar and Maddox also demonstrated 361 W/cm^2 heat removal with two-phase forced convective cooling on an enhanced surface with FC-72 dielectric coolant [1.31].

As discussed by Patel [1.3], liquid cooling is necessary for higher chip powers because of the better thermal transport properties of liquids. Especially in the two-phase regime, latent heat can be used to achieve large amounts of heat removal rates with relatively low chip temperature increase. Considering all possible alternative cooling solutions, micro jets and microchannels are the most promising because of the material and process compatibility with the IC industry, device compactness, as well as the small pumping power required.

1.2 Electroosmotic Micro Coolers

The already commercialized heat pipes have successfully demonstrated that the liquid-vapor phase change can be used in thermal management of individual IC chips. As discussed earlier, because the volume of the liquid contained in a heat pipe system is very limited, it is not expected to work for high heat power applications. However, if an external pump can be used to provide enough pumping energy so that the capillary force is no longer required, the liquid flow rate can be significantly increased, resulting in a huge capacity of heat absorption. The recent development of high-pressure electroosmotic pumps [1.32-1.34] makes it possible to consider an integrated fluidic cooler for the first time. The high-pressure, modest flow characteristics of electroosmotic pumps also drive the design of microchannel coolers to channel diameters below 150 μm.

Figure 1.8 is the schematic of a closed-loop two-phase micro cooler system, invented by researchers at Stanford University. The compact, on-board system consists of a voltage controlled electroosmotic pump, an evaporator and a condenser. The pump moves the working fluid in the liquid phase to the evaporator, which is directly attached to the IC chip. The phase change occurring in the evaporator carries the heat away from the IC chip and the vapor phase moves towards the condenser driven by the pressure difference. The vapor then releases the heat to the condenser and changes back to the liquid phase. Recent studies have demonstrated 38 W heat removal rate with 2 W power consumption [1.35]. The final system is expected to remove 200 W of heat from a single IC chip with less then 5 W of electrical energy.

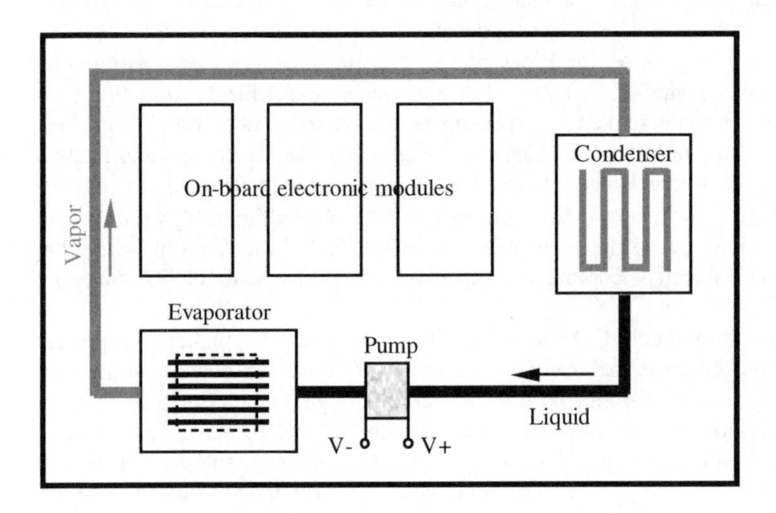

Fig. 1.8 Schematic of a closed-loop two-phase electroosmotic cooler.

From the discussion of the cooling techniques, two-phase cooling with either an enclosed micro jet or microchannel heat sink configuration is the best candidate for the evaporator. Compared with micro jet approach, the two-phase microchannel heat sink is more convenient to implement. This book primarily discusses two-phase flow in microchannel heat sinks. Study of liquid micro jets can be found from related works from Stanford research groups.

1.2.1 Electroosmotic Pumps

The electroosmotic pumps function under the charge double layer that appears at the interface between a liquid phase and a solid surface [1.32-1.34]. As shown in Fig. 1.9, when a liquid is in contact with a solid surface, negative charges appear on the solid wall and attract the positively charged ions in the liquid to form a positive layer very close to the liquid-solid interface. When electrical field exists in the liquid, the positively charged ions tend to move to the cathode and drag the liquid molecules to move in the same direction. When the flow channel is small enough, the drag force can be significant to yield very large pressure loading capability.

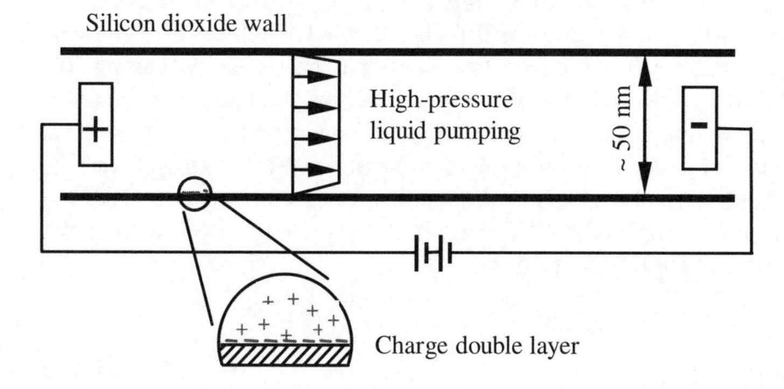

Fig. 1.9 Operating principle of electroosmotic pumps. By courtesy of Juan G. Santiago.

The performance of the electroosmotic pumps is a function of the surface area in a pumping cell, which is currently made of closely packed particles on the order of a few micrometers. Buffered DI (De-Ionized) water is used as the working fluid because it provides plenty of ions to form charge layers; but it is also insulating enough so that a large electrical field can be set across the pumping cell. Taking advantage of micromachining, it is even possible to fabricate silicon electroosmotic pumps and to integrate the pump with the heat sink. Micromachined electroosmotic pumps are currently under development [1.34].

1.2.2 Two-Phase Microchannel Heat Sinks

Microchannel heat sinks have received more and more research interests due to their attractive heat transfer coefficients, which can be increased by a factor of 100 to 1000 with forced two-phase convection compared with forced air convection [1.36-1.41]. These heat sinks have been used in cooling of high-power laser diode arrays, achieving 500 W/cm^2 heat flux removal [1.42-1.44]. Furthermore, boiling convection is even more promising because it requires less pumping power than that of single-phase liquid convection to achieve the same heat sink thermal resistance.

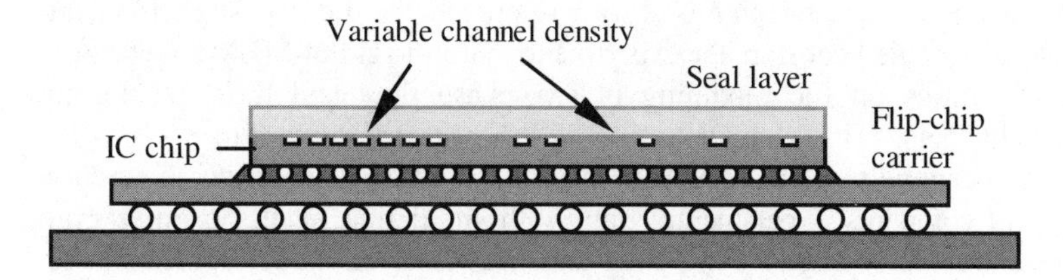

Fig. 1.10 Schematic of a two-phase microchannel heat sink.

Figure 1.10 is a proposed design of the two-phase microchannel heat sink fabricated directly on the back side of a flip-chip bonded IC chip. By removing the thermal epoxy and the lid required for traditional heat sink packaging, the integration of the heat sink with the IC chip significantly reduces the interface resistance between the chip and the heat sink, hence increases the heat removal efficiency. In addition, the microchannels can be defined in groups of varying numbers and geometries, with specially designed channels formed above the hot spots on the chip, for example, the CPU core, hence providing better temperature uniformity across the entire IC chip [1.45].

1.3 Scope of the Book

The closed-loop two-phase microchannel heat sink has the potential to be used in high-power thermal management applications such as individual IC chip cooling. Understanding the two-phase flow behavior in microchannels is the key to successful implementation of such a device. This book focuses on the boiling phenomena in sub-150 μm diameter silicon microchannels, with emphases on the heat transfer experimental system, modeling of two-phase microchannel flows, and the impact of the small dimensions on boiling initiation and flow regimes

The discussion begins with an instrumented microchannel heat sink chip in Chapter 2. By presenting single- and two-phase cooling experiments in microchannels with built-in heaters and temperature sensors, this chapter examines existing controversies in this area, such as whether or not boiling is suppressed in microchannels and whether there is apparent departure of flow behavior in this dimension. Through these preliminary experiments, key problems in this area are identified, including the design of a well-controlled experimental system, modeling of two-phase microchannel flow and heat transfer with supporting experimental data, two-phase flow visualization, and transient measurements. These topics are pursued individually in the following chapters.

Chapter 3 introduces the thermal experimental system that has been used for all experiments discussed in this book. A MEMS device is ideal for microscale thermal experiments by allowing in-situ heat control and temperature sensing. In this chapter, general issues associated with microchannel heat transfer experiments are discussed in detail, including thermal isolation, heat distribution in an experimental system, temperature measurement under heating conditions, and data acquisition systems. This chapter is also a reference for thermal scientists and engineers on microscale heat transfer experiments with the aid of MEMS devices.

Chapter 4 focuses on the modeling of two-phase flow and heat transfer in silicon microchannels. This chapter begins with heat transfer experiments in free-standing microchannels that provide nearly-constant heat flux boundary conditions, and compares experimental data with simulation results from several flow and heat transfer models. The flow and heat transfer model considers the flow regime and predicts local pressure and wall temperatures by applying energy equations to the pre-defined control volumes in the flow channel. Both measurements and modeling of two-phase flow in 28-60 μm diameter channels support that the flow behavior is governed by traditional theories.

Chapter 5 provides visualizations of the bubble formation process in 28-171 μm diameter microchannels and the transient phenomena associated with the nucleation process. This chapter includes sequences of images obtained from high-speed photography in microchannels, which reveals the boiling process as well as two-phase flow conditions. The transient measurement uncovers some unique phenomena, such as transient pressure pulses and the wall superheat during the bubble nucleation process.

Having explored the boiling mechanisms and obtained basic information about two-phase flows in sub-150 μm diameter channels, Chapter 6 explores the reasons for the unique eruption boiling in very small channels. The discussion begins with hypotheses and traditional nucleation theories, and continues to present simulations and experiments that have been specially designed to examine the hypotheses and traditional theories. The conclusion is that the bubble nucleation mechanism does not depart from traditional theories and by using boiling enhancement techniques, the eruption boiling can be avoided.

Chapter 7 summarizes the phase change phenomena in sub-150 μm diameter microchannels and provides general design rules for microchannel heat sinks. A sample design that is capable of removing 200 W from an IC chip with 80 °C temperature increase is provided on the basis of the modeling developed in this study. In the end of the chapter, some interesting issues are addressed for future studies.

2 Two-phase Microchannel Heat Sinks: Problems and Challenges

2.1 Background: Forced Convective Internal Flow

Forced convective internal flow is extensively used in heat exchangers in space heating, power production, and chemical processing. The heat transfer theories for both single- and two-phase internal flows have been well established. However, when the dimension scales down to tens of micrometers in microchannels, due to possible change in dominant forces, the flow behavior can be very different. Questions such as whether or not traditional theories break down need to be answered.

2.1.1 Internal Flow in Macroscale Channels

For viscous internal flow in a macroscale horizontal channel, fluid friction causes a pressure gradient along the flow direction. In single-phase fully developed laminar flow, this pressure drop ΔP can be calculated from

$$\Delta P = \rho \cdot \frac{C}{\mathrm{Re}} \cdot \frac{Lv^2}{2D_H} \tag{2.1}$$

where ρ is the density of the liquid, Re is the Reynolds number, L is the length of the channel, v is the flow velocity, and D_H is the hydraulic diameter of the channel [2.1]. The constant C, equal to 64 for a circular tube, is a factor accounting for the shape of the channel.

When a cold liquid enters a hot channel with a liquid-solid temperature difference, heat transfer between the liquid and the solid occurs. From energy conservation, the heat rate q absorbed by the liquid causes a liquid temperature increase ΔT_l:

$$q = \dot{m} c_p \Delta T_l \tag{2.2}$$

where $\dot{m}$ is the mass flow rate, and c_p is the specific heat of the liquid. When ΔT_l is replaced with the difference between the liquid boiling point temperature and the inlet liquid temperature, Equation (2.2) defines the sensible heat—the maximum amount of heat that can be absorbed by the liquid before the phase change takes place. The opposite is called "latent heat", the heat that is absorbed by the liquid-vapor phase change process without sensible temperature change.

When the liquid temperature becomes higher towards the exit, the wall temperature also increases along the length of the channel. The total heat rate is related to the average wall and liquid temperatures in

$$q = hA(T_{w,avg} - T_{l,avg}) \tag{2.3}$$

where h is a length-averaged convection coefficient, A is the total wetted area, $T_{w,avg}$ and $T_{l,avg}$ are average wall and liquid temperatures respectively.

In detail, the heat transfer from the wall to the liquid results in the development of a thermal boundary layer, in addition to the flow velocity boundary layer. Before the fully developed condition is reached, the fluid temperature, the wall temperature, as well as the local heat transfer coefficient all change as a complicated function of the position; only after the flow is fully developed, the local convection coefficient becomes a constant throughout the channel. There are two boundary conditions that lead to thermally fully developed internal flows—constant heat flux, and constant surface temperature conditions [2.2]. As long as one of the boundary conditions is well maintained, the fully developed flow will appear in the single-phase form and a constant convective coefficient h that is independent of the flow velocity is expected:

$$h = Nu \cdot k_l / D_H \tag{2.4}$$

where Nu is the Nusselt number, and k_l is the thermal conductivity of the liquid. In a fully developed laminar flow, where Reynolds number is below 2300, the Nusselt number is a constant with the value of 3.66 for constant wall temperature and 4.36 for uniform heat flux boundary condition.

The pressure and heat transfer coefficient correlations for two-phase flows are more empirical and strongly depend on the two-phase flow regimes. The major forces that affect the two-phase flow regime include the inertial force, viscous force, interfacial tension forces, buoyancy force, and the exchange of momentum between the liquid and vapor phases. In traditional boiling theory, the phase change process initiates from bubble growth, also called nucleate boiling regime, changes to transition boiling and film boiling with increasing heat flux. The wall temperature is always higher than the boiling liquid temperature so that the heat flux moves from the solid wall to the liquid. It is generally accepted that the wall superheat (the temperature difference between the wall and the liquid during the phase change) is 5-30 °C for nucleate boiling, 30-120 °C for transition boiling, and more than 120 °C for film boiling. Research shows that nucleation sites are normally below 100 μm in size.

2.1.2 Two-phase Flow in Microchannels

The heat transfer can be significantly enhanced when phase change is involved. After Tuckerman and Pease proposed the idea of single-phase silicon microchannel heat sinks [1.19], Bowers and Mudawar first demonstrated that two-phase microchannel heat sinks could remove more than 250 W/cm^2 heat flux with 64 ml/min R-113 flow rate and 20 °C inlet fluid sub-cooling [2.3]. Their research also proved that 510 μm hydraulic diameter channels had smaller thermal resistances compared with 2.54 mm channels, possibly due to thinner thermal boundary layers.

Despite these exciting results, more research must be done in this field before two-phase microchannel heat sinks can be finally commercialized. The major issue lies in the scaling in size. Because traditional fluid mechanics and heat transfer theories are based on studies in macroscale flow channels, when the channel scales down to the dimension barely visible by human eyes, the breakdown of some assumptions can make the well-established theories no longer valid. As an example, in the previous discussion of single-phase flows, viscous laminar flow is an important assumption in the pressure drop and convection coefficient correlations. However, some research indicates abnormal flow behaviors in microchannels. Urbanek et al. showed that the temperature dependence of Poiseuille numbers in 5-25 μm channels was not in good agreement with Navier-Stokes equation predictions [2.4]. Zeighami et al. [2.5] and Tso et al. [2.6] independently reported possible turbulent flow transitions in 100 μm diameter channels at Reynolds numbers below 1600 rather than the theoretical value of 2300. Jacobi further predicted how heat transfer might be affected as the classical fluid model breaks down [2.7].

Another interesting fact is that the nucleation sites on a macroscale surface are usually below 100 μm, which is on the same order of magnitude of the hydraulic diameters of the microchannels. Also in this dimension, the interfacial tension forces become dominant in both the boiling process and the two-phase flow. Thus it is highly possible that the flow behavior as well as the heat transfer in microchannels depart from those in their macroscale counterparts. Many experiments have been carried out on the phase change and the flow behavior in less than 500 μm diameter channels. Rahman et al. fabricated microchannels with KOH wet etch in silicon substrates with doped silicon thermometers to measure the wall temperature at a few points along the test chip [2.8]. Peng et al. investigated flow transition and heat transfer in rectangular microchannels made in stainless steel plates with hydraulic diameters ranging from 200 to 600 μm [2.9]. Jiang et al. designed silicon channels with heaters at the inlet and thermometers on the channel wall to study phase change in diamond-shaped microchannels below 100 μm diameter [2.10]. Many of them have reported unusually high heat fluxes associated with the phase change process in microchannels and proposed new hypotheses to explain how phase change occurs in microchannels.

The previous research provided very good references for the study of two-phase microchannel heat sinks. However, data interpretation in the experiments has been impaired by lack of detailed information about the rate of heat transfer into the fluid and the two-phase flow conditions inside the microchannels. The heat paths in the existing test systems were not examined and the boundary conditions were seldom discussed, hence it is difficult to accurately extract the heat transfer correlations from existing experimental results as a support to more accurate modeling of two-phase heat transfer in microchannels. In addition, there existed inconsistent results and controversies about the nucleation process in microchannels. The question of whether or not typical nucleate boiling occurs in sub-150 μm diameter channels has not been clearly answered. Some researchers reported that bubble nucleation was completely suppressed, while some others did observe phase change, sometimes under unusually large heat fluxes, without knowing how it exactly initiated. If the phase change were suppressed in microchannels, the heat transfer would be made under single-phase only, in which case the heat transfer coefficient would be much lower than that results from the phase change. If this is true, there is obversely no reason to make two-phase heat sinks!

2.2 An Instrumented Microchannel Heat Sink

An instrumented microchannel heat sink has been designed to examine two-phase flow behavior in sub-150 μm channels, with the emphasis on applications in real closed-loop cooling system. A silicon chip with integrated resistive heaters and thermometers is used to emulate a computer chip, with microchannels as cooling elements. The heaters are used to precisely control the heat flux to the chip, emulating the heat dissipated from a working computer chip. Liquid can be introduced to the microchannels to carry the heat away, and the resulting chip temperature is measured with the thermometers. Silicon substrate is selected because the future cooling channels will be formed directly in the IC chips; and the material allows integration of doped silicon resistors on chip. The microchannels are sealed with glass cover slides, which also provide optical access to the internal flow conditions.

2.2.1 Instrumented Heat Sink Design

There have been some previous studies on single- and two-phase microchannel heat sinks [2.11-2.16], where design considerations and modeling were discussed. For simplicity, the design of the instrumented heat sinks begins with macroscale single-phase flow models introduced in Sect. 2.1.1. The design parameters include the pressure drop across the channels and the channel dimensions.

In the proposed closed-loop two-phase cooling system, the electroosmotic pump is expected to provide 5-10 ml/min water flow rate. With this fixed flow rate, Equation (2.1) can be used to determine the target pressure drop and channel dimensions. Equation (2.2) defines the sensible heat when ΔT_l is substituted with 78 °C, the deference between the boiling point of water at atmospheric pressure (100 °C) and the room temperature (22 °C). When combined with Equation (2.3), the average convective coefficient h can be experimentally determined from the measured heat rate, average wall temperature and water temperature increase from the inlet to the outlet.

Both the heaters and the thermometers on the heat sink chip are made of doped silicon resistors. With the resistive heaters, the input heat rate q can be calculated from the product of the input voltage U and current I. The resistance of doped silicon thermometers is a function of the temperature, therefore the wall temperature change can be measured from the resistance variations.

On the basis of trial calculations, three design types were selected for initial experiments. The parameters are listed in Table 2.1. Each chip has an effective cooling area of 1×2 cm or 2×2 cm. The number of channels ranges from 20 to 50, with 50-70 μm channel diameter. Type 1 and Type 3 heat sinks were designed for 10 ml/min flow rate, with larger diameter channels and the maximum pressure drop of around 20 psi. Type 2 was designed for 5 ml/min flow rate with the maximum pressure drop of 54 psi.

Table 2.1 Design parameters of three types of microchannel heat sinks.

	Type 1	Type 2	Type 3
Channel width (μm)	75	50	100
Channel depth (μm)	50	50	50
Channel length (cm)	2	2	2
Number of channels	50	20	30
Effective cooling area (cm^2)	4	2	4
Maximum flow rate (ml/min)	10	5	10
Hydraulic diameter (μm)	60	50	67
Reynolds number*	56	88	79
Pressure drop* (psi)	19.9	53.7	20.1

**The Reynolds number and the pressure drop are calculated under the maximum water flow rate of the design type. Water properties are taken at the room condition (22 ºC, 1 atm).*

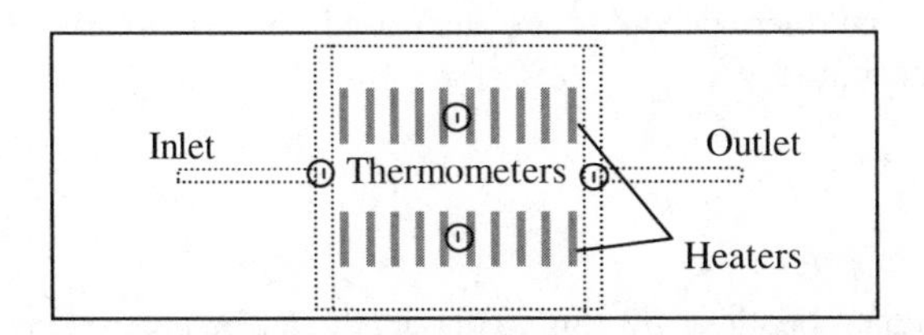

Fig. 2.1 Schematic of the microchannel heat sink chip. The dashed lines outline the microchannels (the middle square) and the inlet/outlet on the front side where the channels are formed. On the back side, twenty parallel heater strips are symmetrically arranged over the cooling area to provide uniform heat flux; and four separate thermometers are located at the entrance, exit and two middle spots to measure local wall temperatures.

Figure 2.1 shows the schematic of the heat sink chip with a detailed layout of the heaters and the thermometers. The overall dimension of the test chip is

2×6.5×0.05 cm, with a same size glass cover slide. The heater consists of twenty resistors connected in parallel, with a total resistance of 100 Ω. Considering the fully developed flow boundary conditions, the resistors are arranged to provide uniform heat flux to the flow channels. Each of the four thermometers has a resistance of 2 kΩ. As the image in Fig. 2.2 shows, the microchannels are defined on the front side of the silicon substrate with a glass cover, and the heaters and thermometers are formed on the back side.

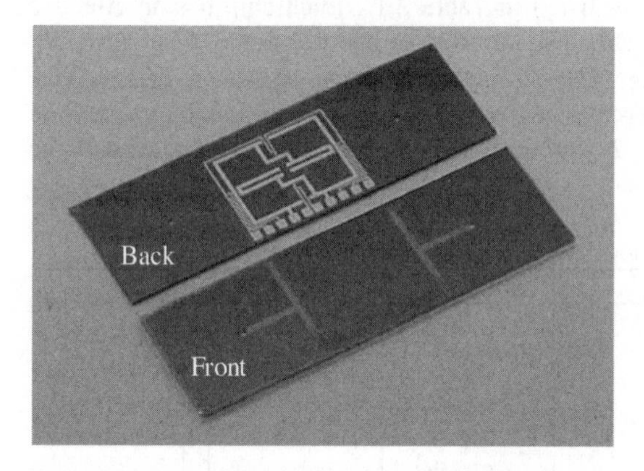

Fig. 2.2 The instrumented heat sink chip. The image shows both the front and back sides of the heat sink chip. On the front side, the two "T" shaped channels are 1 mm wide inlet and outlet; the square between is the effective cooling area, which is 2×2 cm for this device. Doped silicon heaters and thermometers are integrated on the back side.

These instrumented microchannel chips allow accurate control of the input heat rate, local wall temperature measurement, and simultaneous optical observations. External pressure transducers and thermocouples are also used to measure the pressure drop and the fluid temperature.

2.2.2 Device Fabrication

All microchannels have a designed depth of 50 μm. This depth can be precisely controlled by using SOI (Silicon-On-Insulator) wafers that consist of a 50 μm thick silicon layer and a regular support wafer, separated by silicon dioxide. Because of the high selectivity of silicon to silicon dioxide in DRIE (Deep Reactive Ion Etching), the etch rate is very slow on silicon dioxide, which makes the process practically "stops" at the oxide layer, hence the channel depth is defined by the thickness of the SOI.

As shown in Fig. 2.3, the fabrication process begins with the fusion bonding of SOI wafers. The substrate is an N-type <100> wafer of 500 μm thickness. Silicon dioxide of 5000 Å thickness is grown on a pair of one single-side polished wafer

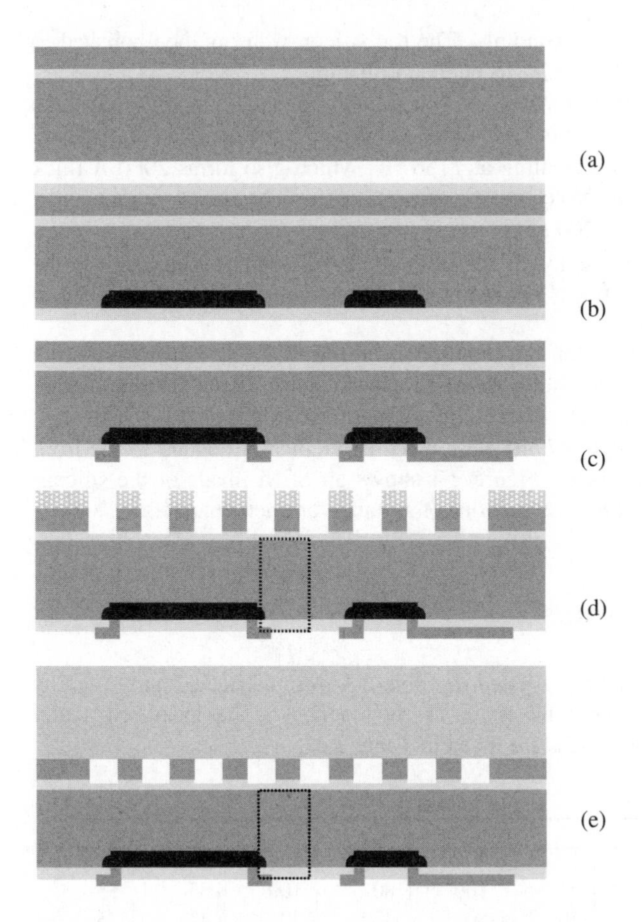

Fig. 2.3 Schematics of the heat sink chip fabrication process. (a) Fusion bonding and polishing of the SOI wafer. (b) Ion implantation and annealing. (c) Aluminum deposition and etching. (d) DRIE on both sides. (e) Silicon-glass bonding.

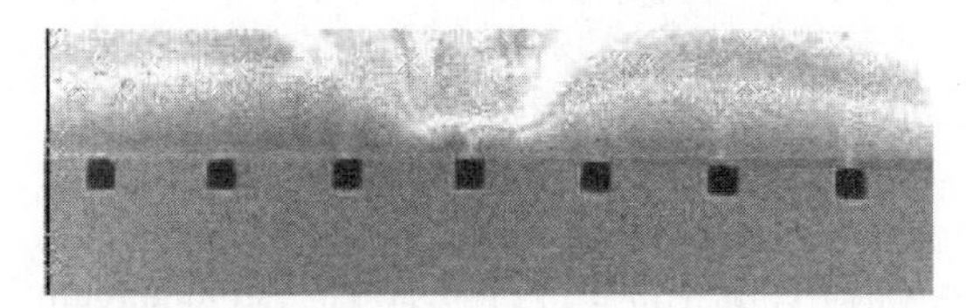

Fig. 2.4 An SEM image of the silicon-glass bonding interface.

and one double-side polished wafer. The single-side wafer is placed on top of the double-side wafer, with the polished side facing down. The two wafers are then temporarily attached by Van der Waals force. An additional 1-hour annealing at

950 °C yields the final fusion bonding. The top silicon wafer is then polished to 50 μm thick with CMP (Chemical Mechanical Polishing).

Second, heaters and thermometers are implanted from the support wafer side (back side). The implant dose is 1E15 Boron at 40 keV energy. The dopants are then activated by a 1-hour annealing at 1150 °C, which also forms 2000 Å thick oxide passivation layer as protection to the resistors. The target surface sheet resistivity after annealing is 200 Ω/□.

Third, 1 μm thick aluminum with 1% silicon is deposited and patterned. In the following aluminum etch, tracks between resistors and contact pads are defined to make electrical connections.

Fourth, DRIE is used to etch the channels from the SOI side, and then to etch the two inlet/outlet through holes from the back side, shown as the dashed rectangle in Fig. 2.3(d) and (e). Both etches stop at the oxide layer.

Last, a 500 μm thick Pyrex 7740 glass piece is anodically bonded to the front side to seal the microchannels. Figure 2.4 shows an SEM image of the silicon-glass bonding interface with cross sections of rectangular microchannels.

2.3 Experimental Results

Six heat sink chips of three types listed in Table 2.1 were fabricated and tested by applying heat rate and using DI water as the coolant. The measured actual dimensions of the testing devices are listed in Table 2.2.

Table 2.2 Measured dimensions of test heat sinks.

Device number	1	3	4	5	6	7
Type	1	3	1	3	2	2
Channel width (μm)	75	100	80	100	55	55
Channel depth (μm)	50	50	42	42	42	45
Channel length (cm)	2	2	2	2	2	2
Number of channels	50	30	50	30	20	20
Hydraulic diameter (μm)	60	67	55	59	48	49

2.3.1 Experimental System

The experimental system, shown in Fig. 2.5, allows DI water to be pumped to the channels at a constant flow rate and a constant power to be applied to the on-chip heaters to emulate a working computer chip with heat dissipation. An aluminum fixture is made to mount and support the heat sink chip. The fixture has internal flow directors and O-rings to securely seal the inlet and outlet holes in the microchannel chip. A syringe pump supplies DI water to the system. A pressure transducer and a thermocouple measure the inlet pressure and the exit fluid temperature respectively. The heat rate, wall temperatures, the pressure drop along the channels, and the outlet fluid temperatures are recorded simultaneously.

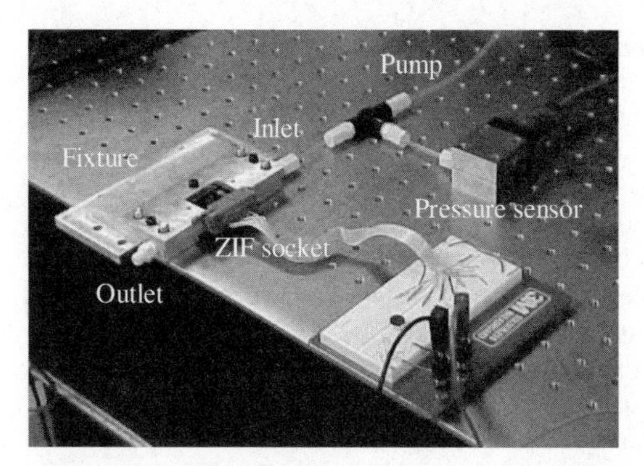

Fig. 2.5 Experimental system. The heat sink chip is supported and sealed by the fixture with internal flow directors. The ZIF socket connects the heaters and thermometers to the breadboard, where input power is supplied and the thermometer resistances are measured. A pressure transducer is attached at the inlet. The outlet fluid temperature is measured with a thermocouple.

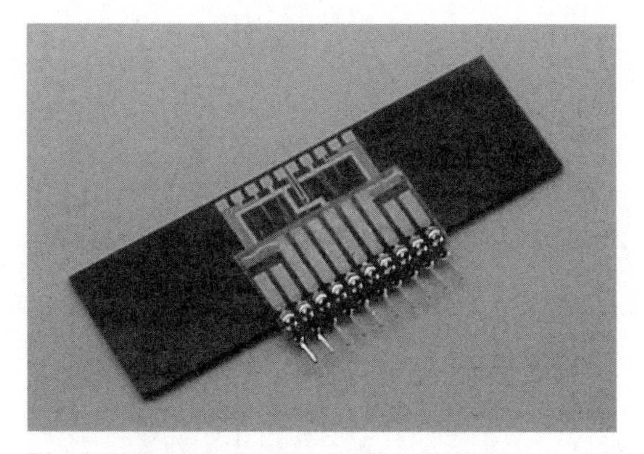

Fig. 2.6 The wire-bonded Surfboard (surface-mounting board) and the microchannel chip. The Surfboard is attached to the chip with epoxy, and the aluminum pads on the chip are wire-bonded to the Surfboard.

Because of the dimensions of the chip and the nature of the experiment, the chips cannot be packaged in traditional ceramic packages with bond pads. Here a Surfboard™ (the breadboarding medium for surface mount components by Capital Advanced Technologies) is used to make electrical connections to the device chip. As shown in Fig. 2.6, the Surfboard is attached to the resistor side of the chip, and the aluminum pads are wire-bonded to the Surfboard, which has ten standard single-in-line pins. After the microchannel chip is mounted to the fixture, a ZIF (Zero-Insertion-Force) socket with ribbon wires is clamped to the Surfboard to make further wiring and connections.

2.3.2 Thermometer Calibration

Doped silicon thermometers on the same chip usually have similar but not identical resistances and thermal coefficients. For the highest accuracy, each thermometer on all the devices was calibrated individually prior to heat transfer measurements. During calibration, the device chip was placed in a convection oven with uniform temperature field. The reference temperature from a calibrated thermocouple and the resistances of the thermometers were recorded at 5-10 °C increments from room temperature to 180 °C. Figure 2.7 is the four thermometer calibration curves from one heat sink chip.

After recording the temperature response of the resistors, each of the calibration curves is plotted and a polynomial fitting equation is calculated as shown in Fig. 2.8. In the later experiments, the resistance variations of the thermometers are converted to temperature data from the temperature-resistance equations. Since the major source of error comes from the temperature calibration and data fitting, the accuracy of the measurement is estimated from the mean square deviation in the data fitting, which is ±2 °C for these thermometers.

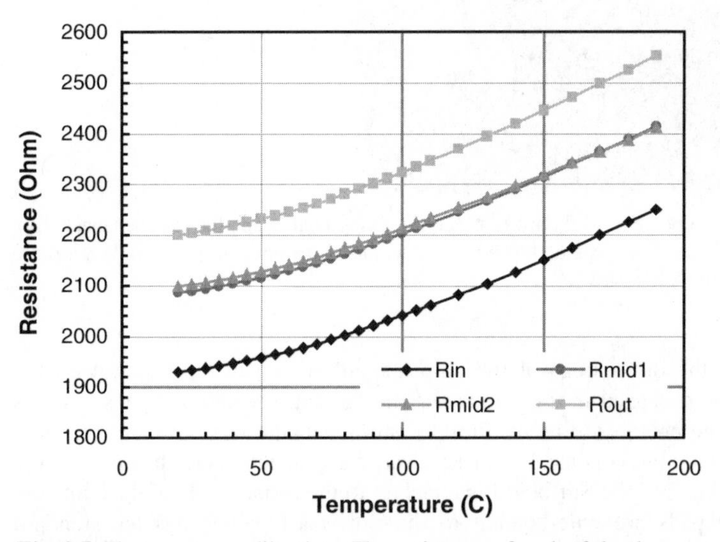

Fig. 2.7 Thermometer calibration. The resistance of each of the thermometers is recorded at various temperatures. A uniform temperature field convection oven is used for the calibration, and a thermocouple provides the reference temperature.

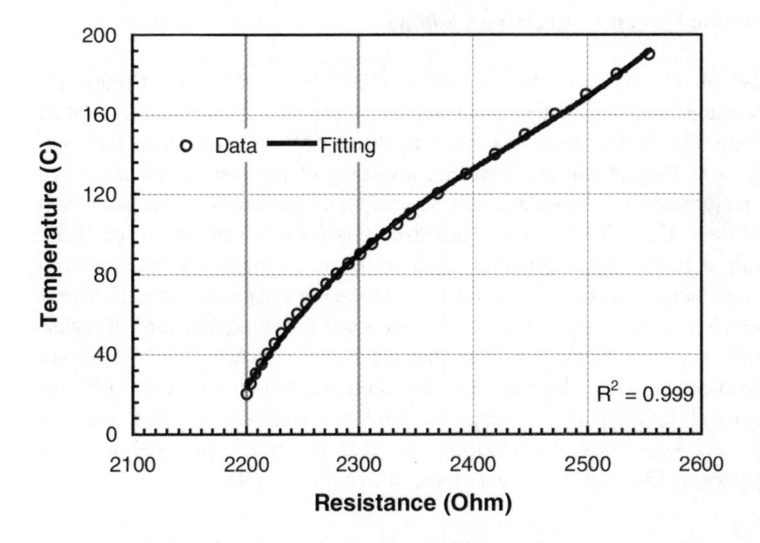

Fig. 2.8 The fitting curve of a typical thermometer. The polynomial fitting curve is used as the temperature-resistance function of this thermometer. Repeat calibrations on resistors after use indicated variations of less than 0.2%.

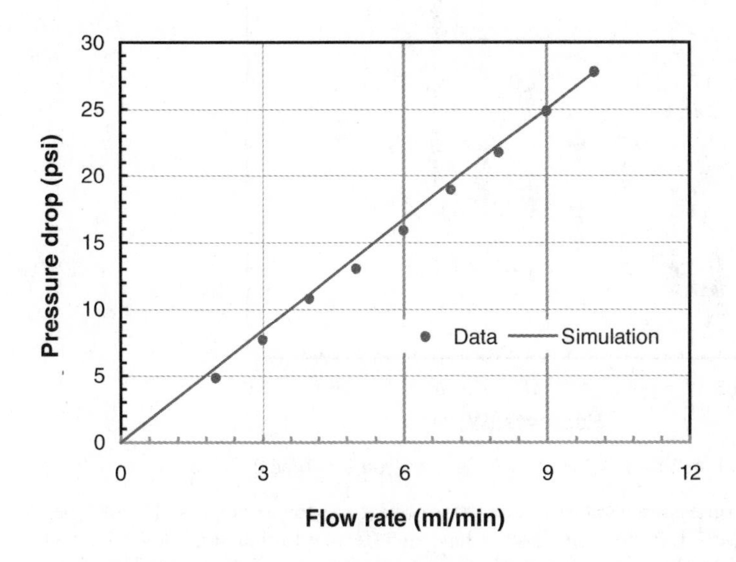

Fig. 2.9 Pressure drop along the microchannels of device No. 4, a Type 1 device (each channel has 55 μm hydraulic diameter). This figure shows the measured and simulated pressure drop along the microchannels as a function of water flow rate at room temperature. A simple laminar flow model described in Equation (2.1) with constant friction factor accurately describes this data.

2.3.3 Single-phase Forced Convective Cooling

The first phase of the experiments was single-phase measurements. Figure 2.9 shows the measured and simulated pressure drops in the microchannels against flow rates. The noise in the pressure sensor measurements was less than 0.05% of the full scale, or 0.05 psi; and the absolute accuracy of the sensor was ±0.1 psi according to manufacturer data sheets. These lead to a total pressure measurement uncertainty of less than 0.15 psi. The simulation model is based on fully developed laminar flow with a constant friction factor, as expressed in Equation (2.1). The shape factor C is defined as 62 for rectangular channels with 2:1 aspect ratio. The measured dimensions in Table 2.2 are used in the simulation. Because Reynolds numbers are all below 100, the internal flow is expected to be laminar. The good agreement between the measurement data and the simulations indicates that the traditional laminar flow model is valid for channels on the order of 100 μm with very small Reynolds numbers. This result is also supported by experiments performed by Lee et al. [2.17] and Judy et al. [2.18].

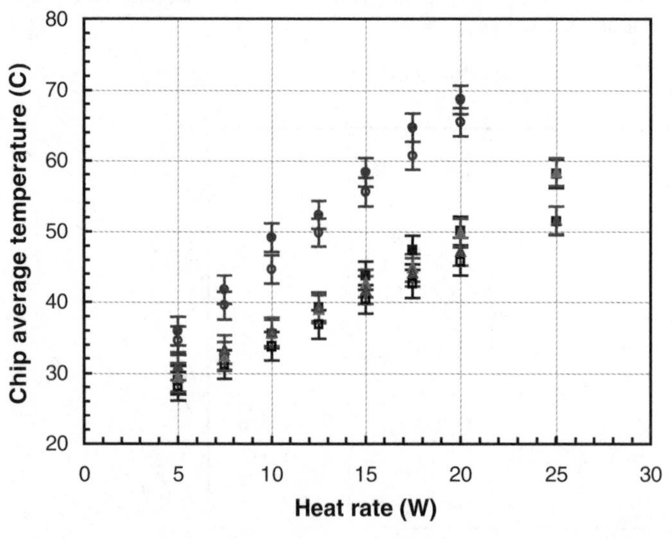

Fig. 2.10 Chip average temperature as a function of heat rate for six devices of three types. As shown in Table 2.1, Type 1 and Type 3 have an effective cooling area of 4 cm^2, with different size and number of microchannels, but similar entire convection area; Type 2 has an effective cooling area of 2 cm^2. Type 1 and Type 3 devices were tested at 10 ml/min flow rate, and Type 2 at 5 ml/min. Since Type 2 has a lower flow rate and less cooling area, the chip temperature rise is higher than Type 1 and Type 3 at the same heat rate. All trend lines extrapolate to the room temperature 22 °C for zero heat input.

The chip temperature as a function of increasing heat rate at steady state was also measured for all six devices under single-phase flow regime, as shown in Fig. 2.10. In accordance with Equations (2.3) and (2.4), at a constant mass flow rate, the average chip temperature linearly increases with the heat rate under single-phase forced convective flow. The slopes of the curves represent the thermal resistance of the heat sinks: at 5-10 ml/min water flow rate, 25 W of heat is removed with 50 °C average chip temperature rise, yielding approximately 1.3-2.3 °C/W thermal resistance. From Equation (2.3), the average heat transfer coefficient h is found to vary in the range of 8000-15000 W/m^2-K. These initial single-phase forced internal flow experiments do not indicate any departures from macroscopic behavior.

2.3.4 Two-phase Forced Convective Cooling

The second phase of the experiments was to explore the phase change in the microchannel heat sinks. Theoretically, phase change occurs when the input heat rate exceeds the sensible heat of the liquid. As a first attempt, the input heat rate was fixed at 25 W, and the flow rate was reduced from 10 ml/min to 1 ml/min in Device No. 4, with 50 channels of 55 μm in hydraulic diameter. The average chip temperature change as a function of water flow rate at steady states is plotted in Fig. 2.11. Same as in Fig. 2.10, the chip temperature varies approximately linearly with the inverse of the flow rate under the single-phase flow regime. When the flow rate was further reduced to 3 ml/min, boiling was observed in the channels.

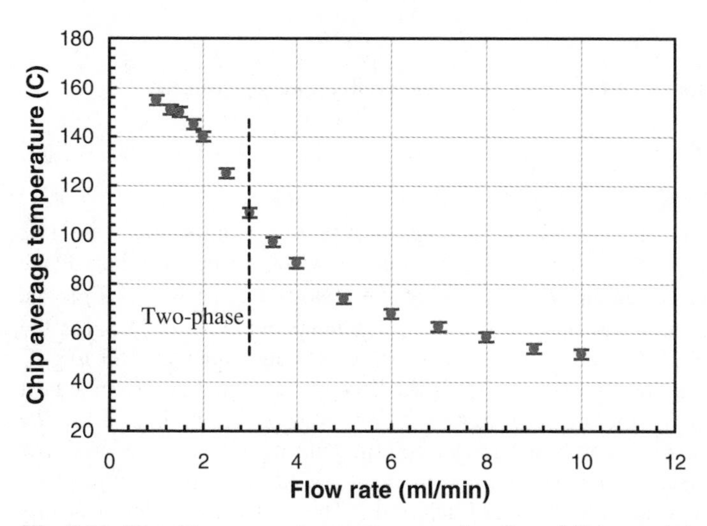

Fig. 2.11 The chip average temperature as a function of flow rate below 10 ml/min for a constant heating power of 25 W. Device No. 4 of Type 1 (each channel has 55 μm hydraulic diameter) was tested as flow rate was reduced from 10 ml/min to lower values, and boiling was observed at flow rates below 3 ml/min.

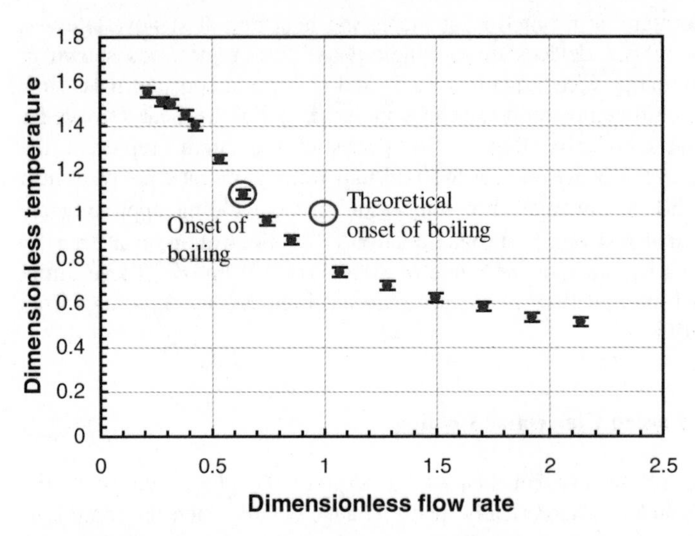

Fig. 2.12 Dimensionless plot of the chip average temperature change as a function of the flow rate in Fig. 2.11.

In the dimensionless plot in Fig. 2.12, the variables are replaced with

$$T^* = \frac{T_{avg}}{T_{sat}}, \text{ and } Q^* = \frac{Q}{q / \rho c_p (T_{sat} - T_0)} \tag{2.5}$$

where T_{avg} and Q are the measured chip average temperature and flow rate respectively. T_{sat} is the saturation temperature of water. Since the exit of the channels are open to the air, T_{sat}=100 °C at the atmospheric pressure. T_0 is the inlet water temperature, assumed to be the room temperature of 22 °C; q is the heat rate, equal to 25 W; and c_p is the specific heat of water, equal to 4185 J/kg-K. The bottom term in the expression of Q^* is the flow rate at which the sensible heat of water is equal to the input heat rate 25 W. Therefore, in Fig. 2.12, the coordinate (1,1) is the theoretical onset of boiling point with the fixed 25 W input heat rate. The offset of the measured boiling point indicates that the absorbed heat by the water is less than the measured heat rate. This is because of the heat loss due to convection and radiation from the chip surface and conduction into the aluminum fixture. However, this experiment proves that phase change does occur in these silicon microchannels.

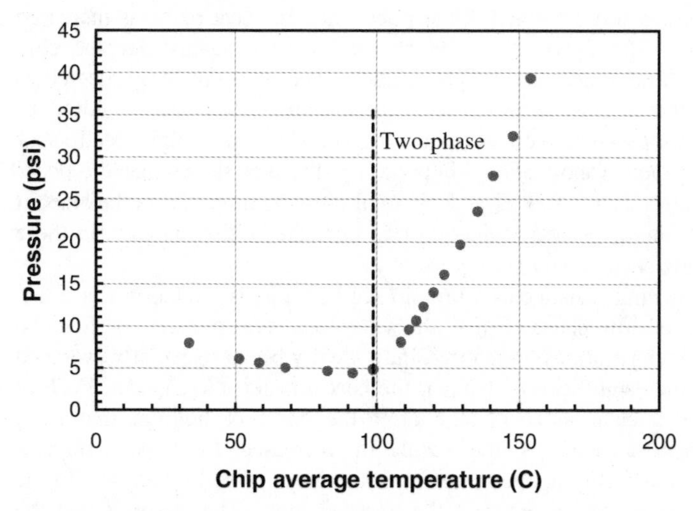

Fig. 2.13 Inlet pressure change from single-phase to two-phase flow. Device No. 4 was tested at constant flow rate of 1 ml/min with varying heat rate. The pressure drop along the microchannels can vary dramatically because of the property difference between water and vapor.

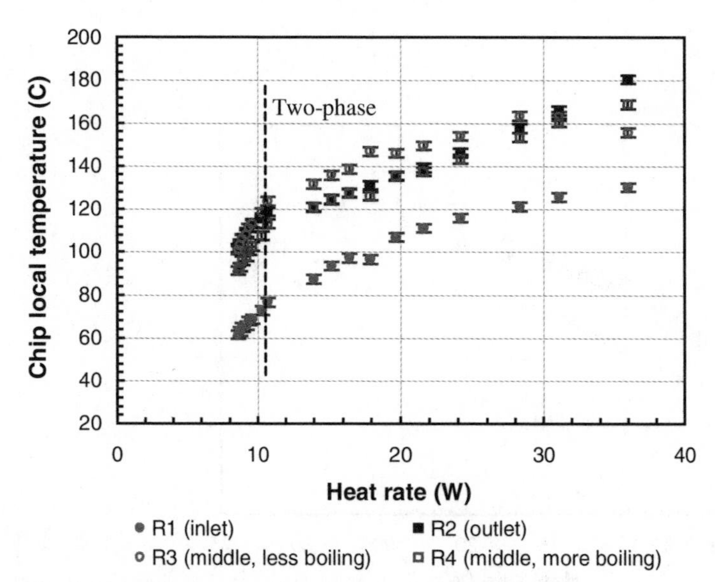

Fig. 2.14 Boiling curves of Device No. 4 at a constant DI water flow rate of 1 ml/min with increasing heat rate at steady states.

Understanding that the phase change would occur in this heat sink chip, two-phase experiments were carried out with a constant flow rate of 1 ml/min.

Figure 2.13 is the pressure recorded at the inlet when the heat rate was increased from 0 to 35 W. The figure plots the pressure change against average chip temperature. In the single-phase flow region, the pressure drop decreases corresponding to the decrease of water viscosity at higher temperatures. After the onset of boiling, the pressure drop increases dramatically due to high-speed vapor flow generated by rapid vaporization. Importantly, the plot shows that the phase change occurs slightly above 100 °C at 5 psi total pressure drop. From the amount of wall superheat, this is a good indication that nucleate boiling happens in these 55 μm hydraulic diameter channels.

Local wall temperature change as a function of heat rate is plotted in Fig. 2.14. For a better sense of how phase change alters the local heat transfer coefficients, the four measured temperatures are plotted individually because spatially averaged wall temperature may significantly dampen the heat transfer rate change. Each of the curves shows a clear slope change after the onset of boiling, that is, a temperature plateau appears as the result of increased local heat transfer coefficient. The higher chip temperatures after the onset of boiling are due to the increase in the pressure associated with the phase change. The reason is that the boiling point of water is a strong function of pressure—at the maximum gauge pressure of 40 psi, the boiling point of water shifts to 152 °C, which explains the high temperatures at large heat rates. Partial dry-out in the channel is also responsible for the extremely high wall temperatures.

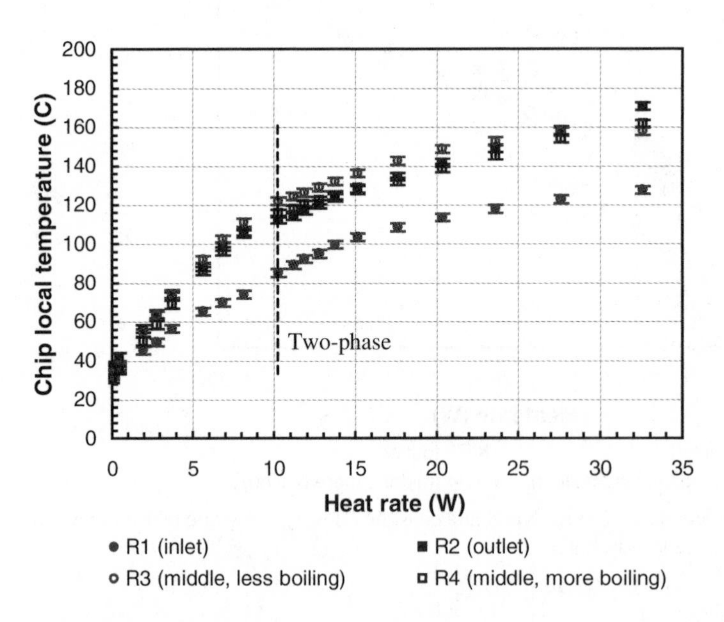

Fig. 2.15 Boiling curves of Device No. 4 at a constant DI water flow rate of 1 ml/min with decreasing heat rate at steady states.

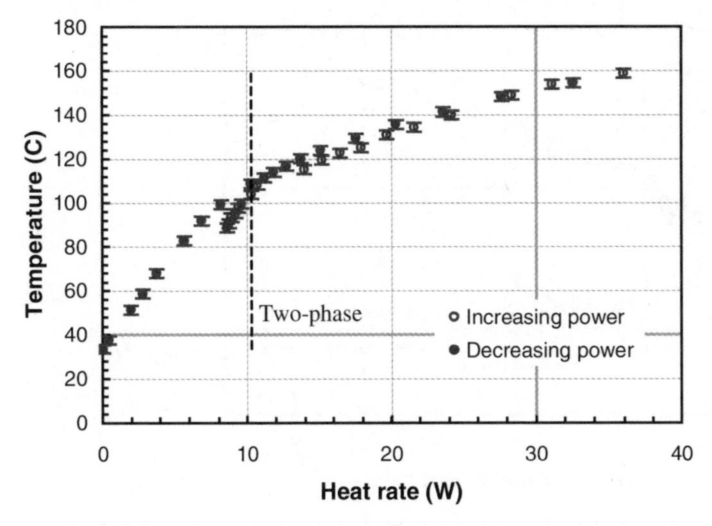

Fig. 2.16 A complete boiling curve with increasing and decreasing power cycles from one of the middle thermometers on Device No. 4 at constant DI water flow rate of 1 ml/min.

The temperature change as a function of decreasing heat rate was also measured and plotted in Fig. 2.15. The data show the same trends in the single-phase and two-phase regions without significant hysteresis. Because of the warming-up of the fixture due to the heat conduction, the chip temperature at zero heat rate is slightly higher than the room temperature of 22 °C.

Figure 2.16 shows a complete boiling curve measured from a middle thermometer on Device No. 4 in one power cycle. In summary, in the single-phase region, the wall temperature linearly increases with increasing heat rate as expected. In the two-phase region, a temperature plateau appears because of the latent heat. Comparing the thermal resistances, at 1 ml/min water flow rate, the heat sink yields approximately 10 °C/W in the single-phase region and 2 °C/W in the two-phase region. With very low liquid flow rate of 1 ml/min, it is not surprising that the numbers look disappointing compared with the industrial expectation of 0.1 °C/W. However, the thermal resistance is lowered 5 times just by introducing the phase change.

In dimensionless form, the temperature and the heat are expressed with

$$T^* = \frac{T_{avg}}{T_{sat}}, \text{ and } q^* = \frac{q}{\dot{m}c_p(T_{sat} - T_0)} \tag{2.6}$$

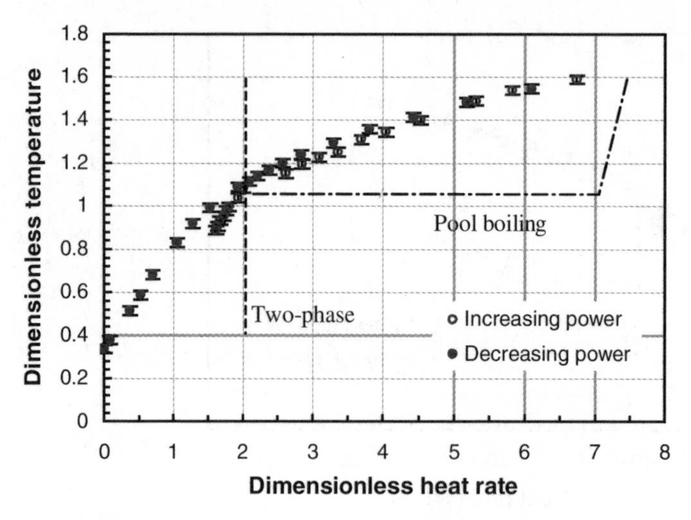

Fig. 2.17 Dimensionless plot of the local wall temperature change as a function of the heat rate.

where the wall temperature is divided by the water saturation temperature and the heat rate is divided by the sensible heat of a given flow rate $\dot{m}$. In Fig. 2.17, the region with dimensionless temperature greater than 1 is the theoretical two-phase region. Comparing with classical pool boiling curves shown in dashed lines, forced convective boiling in microchannels has a larger temperature-heat rate slop, or thermal resistance, due to the potential increase in the pressure. This effect must be considered in the microchannel heat sink design with the help of an accurate two-phase flow model.

2.4 Problems and Challenges

The preliminary experiments on the two-phase internal flow have successfully demonstrated two-phase cooling in microchannel heat sinks. The instrumented heat sink chips with integrated semiconductor heaters and thermometers significantly help the study of two-phase convection in microchannels. Phase change has been observed in 55 μm diameter channels, and the experimental data indicate that phase change occurs as typical nucleate boiling with less than 30 °C wall superheat. The superheat is calculated from the difference between the chip temperature and the boiling point of water at the measured pressure. These experiments, along with existing research in this area, address some problems and challenges in the study of two-phase heat transfer.

First, two-phase flow and heat transfer models are the key references in microchannel heat sink design; hence a valid model with supporting data is necessary in this field. Traditional two-phase heat transfer models are mostly supported by empirical correlations. Two important parameters have been traditionally used, namely, critical heat flux (the maximum heat flux for pool boiling before liquid dry-out condition appears) and convective coefficient. However, the nonuniformity nature of forced convective cooling imposes complexities to both concepts. On the one hand, the definition of "critical heat flux" is no longer a well-represented concept. In two-phase microchannel flows, the appearance of dry-out spots is not only a function of the heat flux, but also closely related to the flow rate, flow regimes, pressure, channel dimensions, and many other factors. On the other hand, the two-phase forced internal flow in microchannels combines the features of single-phase forced convection, two-phase heat transfer, and possibly unusual behavior due to the scaling effect, which makes its local heat transfer coefficients even more complicated. Therefore, the only possible means of study is to build flow and heat transfer models on the basis of experimental data. This requires precise control of the experimental conditions, including the heat distribution in the entire experimental system.

Fig. 2.18 An image of water boiling in the microchannels with a hydraulic diameter of 55 μm. The shiny lines in the cooling area are where the bubbles form and move through the chip from the inlet to the outlet. Since there is a temperature gradient in the silicon substrate, the entire cooling area does not boil at the same time. Bubbles generate at the exit manifold of channels first, then the origin of the bubble line gradually moves to the entrance manifold in accordance with the local substrate temperature rise.

Second, a well-defined heat transfer experiment requires precise control of the heat rate and accurate measurement of local wall temperatures. The wall temperature obtained from a heat sink design as described in this chapter as well as in existing research can only provide a spatially averaged value from a group of channels, which does not exactly represent the local heat transfer under a specific flow regime. For example, Fig. 2.18 is an image of water boiling in Device No. 4, microchannels of 55 μm hydraulic diameter. The nonuniform boiling across the channels may be caused by some local wall surface conditions. When this kind of nonuniform boiling happens, the averaged wall temperature measurement does not provide much information on the local heat transfer conditions in channels with a two-phase flow. Besides, it is difficult to evenly distribute liquids into a large group of channels from a common manifold or reservoir. The uneven flow field adds more uncertainties in the control of experimental conditions. These problems call for a well-designed microchannel device that allows more accurate heat transfer experiments, for example, an instrumented device with a single test channel or a small number of microchannels that minimizes the flow interference between channels.

Third, flow regime visualization is the key to understanding the two-phase flows in microchannels. Since the heat transfer is dictated by the flow regime, the flow patterns are the most important reference in selecting the two-phase flow models. Although the current experimental data seem to prove that the wall superheat is within the nucleate boiling regime, only visualization of the boiling process can directly answer the question whether or not the two-phase flow in microchannels depart from its macroscopic behavior.

Forth, transient measurements will help understand the phase change process in microchannels. Variations in the resistances of the thermometers after the onset of boiling have been observed in all two-phase experiments, as a strong comparison of steady resistance readouts in the single-phase regime. Temperature variations of up to 3 °C at frequencies of 10 Hz and above have been measured, indicating that dynamic thermal development are possible in these structures. Also, local chip temperature fluctuations of about 5 °C have been noticed at the outlet when the bubbles form and pass the channels. These phenomena have not been addressed by existing research and are unique in microchannels, hence further transient measurements will be valuable to understanding the phase change process.

The ultimate question in the study of forced two-phase microchannel flow and heat transfer is whether or not traditional theories break down in this dimension. In the following chapters, the topics addressed here are individually explored, leading to the final answer to the question.

3 A Thermal Experimental System with Freestanding Microchannels

3.1 Thermal Isolation of Microchannels—Design Concept

MEMS technology can significantly facilitate the study of two-phase flow in microchannels by providing instrumented chips with multiple control and measurement accesses. The importance of heat distribution control in the experimental system as well as the thermal isolation of the test channel has been addressed in the last chapter. The discussions lead to the idea of experimental devices with a single channel or a very small number of channels for accurate heat transfer measurement. With an instrumented MEMS device, the only possible problem lies in the excellent thermal conduction in the silicon substrate. Having a thermal conductivity of k=148 W/m-K, silicon is a better thermal conductor than most metals. This helps to distribute heat from a computer chip to its heat sink, but in thermal experiments, it threatens the accuracy of local wall temperature measurements, because the heat can easily diffuse from high temperature areas to low temperature areas. To minimize the heat diffusion problem, FEM (Finite Element Modeling) has been used to simulate the thermal conduction in a silicon substrate to design the geometry of the single-channel devices.

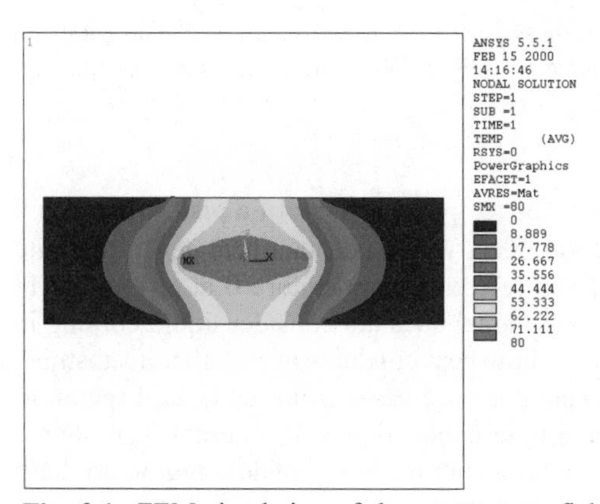

Fig. 3.1 FEM simulation of the temperature field of a silicon chip heated at the center region. As suggested by a conduction model with internal heat source, the temperature field is parabolic from the center to the edges.

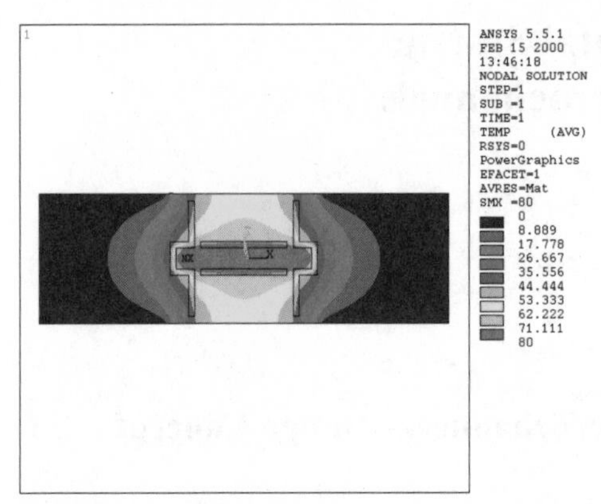

Fig. 3.2 FEM simulation of the temperature field of a silicon chip with through-chip trenches around the heat source. The temperature field has a modified "parabolic" profile, with a plateau in the region confined by the trenches.

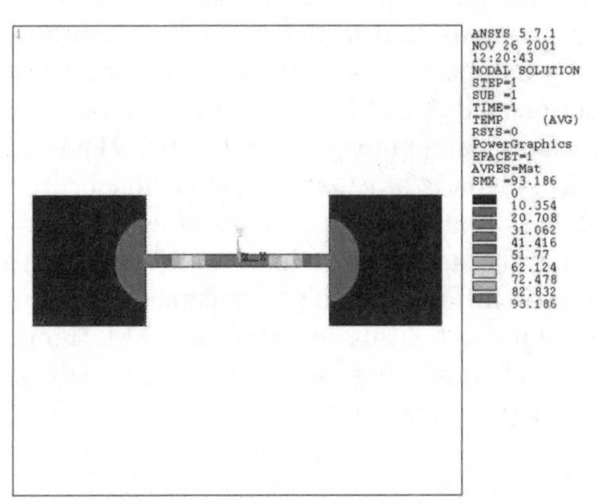

Fig. 3.3 FEM simulation of the temperature field of a silicon chip with heat source defined on a freestanding beam. The heat source is well isolated due to the large conduction resistances to the sides of the chip.

Figure 3.1 is the FEM simulation of the temperature field of a plain silicon chip of 6.5 cm long, 2 cm wide, and 500 μm thick, with a constant heat generation rate of 0.4 W at the center region and fixed reference temperature 0 at left and right edges. Although this simple conduction model does not consider liquid cooling in microchannels, it provides some idea of how heat conducts in the silicon substrate. The simulation results show that temperature decreases from the heated region to the edge of chip as suggested by a simple conduction with internal heat source solution. However the top and bottom temperatures in the middle region are both high, indicating large conduction flux in this area.

In a slightly modified structure, as shown in Fig. 3.2, a few through-chip trenches are added to isolate the heater region. As expected, here the high-temperature region is practically confined by the trenches due to the reduced thermal conducting area.

It is then natural to postulate that a freestanding beam is ideal for thermal isolation of the heater because the limited dimensions of the beam minimize heat conduction paths to the substrate. This is confirmed by the FEM modeling shown in Fig. 3.3. With these simulations, a freestanding beam structure is selected to minimize the heat conduction in the silicon substrate for single-channel devices.

3.2 Heat Distribution—Thermal Resistance Model

With the help of FEM simulations, the new microchannels are designed on a freestanding beam with integrated heaters and thermometers. This concept is shown in Fig. 3.4. The test channel lies on a freestanding beam, with large reservoirs at the two ends. In the realistic experimental system, the heat flux can be conducted in silicon and raise the liquid temperature before it enters the microchannel, causing uncertainties in the inlet liquid temperature. However, if the volume of the reservoir is big enough and the "leakage" heat flux is negligible, pre-heating of the liquid can be neglected. That is, the two large reservoirs further provide heat insulation between the channel and the substrate. In addition, the test channel extends slightly beyond the heater length as shown in Fig. 3.4, so that the heater is not directly heating the reservoir. These two insulating regions are referred to as the "entrance region" and the "exit region".

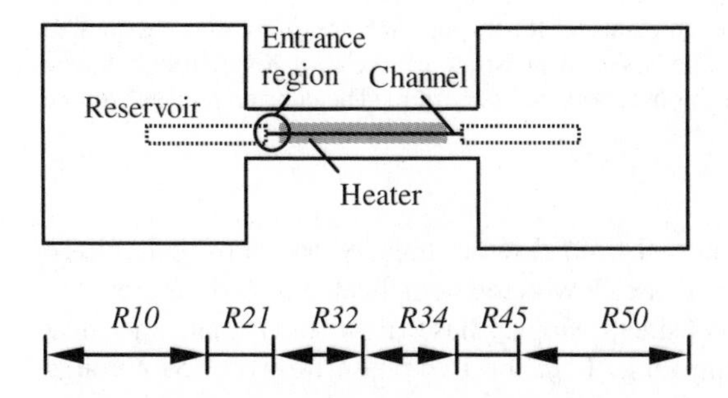

Fig. 3.4 Schematic of the single channel with heaters on a freestanding beam design. The definitions of conductive thermal resistances R10-R50 are also given, which will be used in Fig. 3.5 and further discussions.

The heat paths in the experimental system must be well defined so that the amount of heat loss in the test system can be predicted or measured. Heat loss here is defined as the fraction of the input heat rate that is carried away by any means other than convection to the fluid. For instance, heat conduction to the fixture and free convection from the silicon chip are both considered as heat loss.

A thermal circuit has been developed to simulate the heat distribution in the system. In a thermal circuit, heat paths including conduction, convection and radiation are modeled as electronic resistances; the heat rate is modeled as a power source; and heat fluxes as well as temperatures are represented by currents and voltages in the thermal circuit. The current distribution in a thermal circuit is equivalent to the heat distribution in a thermal system.

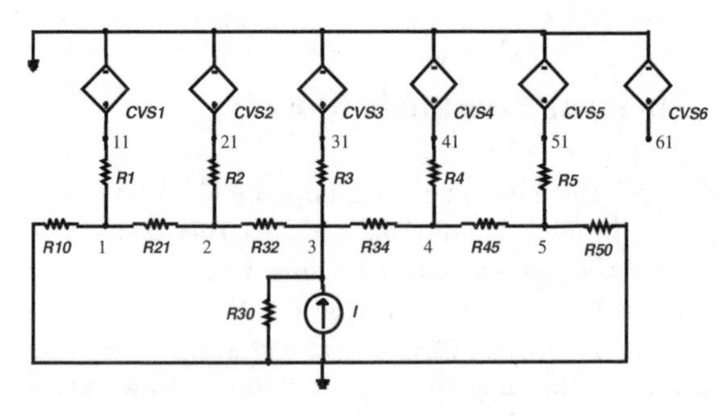

$$CVS1 = I(R1) * Rw/2$$
$$CVS2 = CVS1 + I(R1) * Rw/2 + I(R2) * Rw/2$$
$$CVS3 = CVS2 + I(R2) * Rw/2 + I(R3) * Rw/2$$
$$CVS4 = CVS3 + I(R3) * Rw/2 + I(R4) * Rw/2$$
$$CVS5 = CVS4 + I(R4) * Rw/2 + I(R5) * Rw/2$$
$$CVS6 = CVS5 + I(R5) * Rw/2$$

Fig. 3.5 Schematic of the thermal circuit for the freestanding beam design. The controlled voltage sources (CVS) simulate the fluid, whose temperature rise depends on the heat absorption rate and the liquid entry temperature; R1 through R5 are convective resistances from silicon wall to the fluid; all other thermal resistors are the heat loss through silicon substrate, including conduction, free convection, and radiation. The constant current source I represents the constant input heat rate.

As shown in Fig. 3.5, the internal fluid flow is modeled as controlled voltage sources (*CVS*). Because the two-phase flow is too complicated to be interpreted in thermal circuits, the model considers single-phase flow only; and the heat distribution is assumed to be the same as in the two-phase flow. *CVS1* through *CVS5* simulate the forced convective heat transfer in the inlet reservoir, entrance region, microchannel, exit region, and the outlet reservoir; and *CVS6* is used to estimate the exit liquid temperature at Node *61*. R_w is the equivalent resistance of convective flow with heat transfer ($1/\dot{m}\, c_p$, where $\dot{m}$ is the mass flow rate, and c_p is the liquid specific heat). The thermal implication of the *CVS* is that the average fluid temperature in one flow segment is equal to the sum of the inlet temperature and a half of the temperature rise in this segment. The constant input heat power is modeled as a constant current source. The ground in the circuit represents the room temperature 22 °C.

In the thermal circuit, Nodes *1* through *5* are the average temperatures of the inlet reservoir, entrance region, microchannel, exit region, and the outlet reservoir. Nodes *11* through *51* are the average temperatures of the fluid in the corresponding flow segments. Node *61* is the exit fluid temperature. *R1* to *R5* are forced convective resistances (*1/hA*, where *h* is the convective coefficient, and *A* is the wetted area); the rest of conductive resistances *R10* through *R50* are defined in Fig. 3.4, where *R10* and *R50* represent the sum of device-system contact resistance, system conductive, free convective and radiative resistances; *R30* is free convective and radiative resistance of the device chip; *R21*, *R32*, and *R34* are conductive resistances in silicon (*l/kA*, where *l* is the length, *k* is the thermal conductivity of silicon, and *A* is the cross sectional area).

Ideally, the total heat rate, which is modeled as the current source *I*, should go through *R3*, representing the forced convection in the microchannel, and cause fluid temperature rise. However, heat loss always exists due to the existence of alternative heat paths, which are represented by the currents in other resistors in the circuit. By simulating the currents in all resistors, the heat distribution in the system can be predicted.

Table 3.1 Thermal resistances of a freestanding microchannel and a heat sink chip.

Dimension and resistance	Freestanding beam	Heat sink
Channel wall	20×0.5×0.5 mm	20×20×0.5 mm
Reservoir	0.65×0.25×10 mm	1×0.1×10 mm
R_w ($1/\dot{m}\, c_p$, fluid)*	143.5 K/W	143.5 K/W
R1/R5 (*1/hA*, reservoir)*	13.8 K/W	3.3 K/W
R21/R45 (*l/kA*, channel)	80 K/W	2 K/W
R32/R34 (*l/kA*, channel)	300 K/W	7.5 K/W

* *Assuming 0.1 ml/min water flow rate.*

The thermal circuit analysis indicates that the system heat loss due to conduction in silicon can be limited by adjusting the ratio of *R3* to other resistors, that is, maximize thermal resistances other than *R3*. This can be achieved by adjusting the geometry of the device chip. For instance, by reducing the conduction in silicon and using a deep, large reservoir help thermally isolate the microchannel, so that the pre-heating of the liquid in the inlet manifold is minimized. The freestanding beam should be as narrow as possible to reduce the heat loss through conduction in silicon substrate because the thermal resistance is inversely proportional to the conducting area. As an example, with 0.5 mm beam width, the conduction resistances of a freestanding beam design are compared with the previous heat sink chip in Table 3.1. With much larger thermal resistance in the channel walls, the conduction in the bottom of the microchannel is significantly reduced; therefore much more accurate local wall temperature measurement is expected for channels formed on a freestanding beam. In addition, the uniform heat flux boundary condition is enhanced by this design because the heat is deposited to a nearly isolated beam.

The model also suggests that varying flow rates result in varying heat distribution in the system because all convective resistances (*R1* through *R5)* are functions of the flow rate. This change can be significant, so the heat loss in the system must be estimated for each experimental condition. The selection of design parameters depends upon the detailed requirements of specific experiments, which will be discussed in individual chapters.

3.3 Thermometry—Wall Temperature Measurement

One characteristic of this microchannel chip is that the heaters and thermometers must be integrated onto a very narrow, freestanding beam in order to locally deposit heat into the flow channel as well as to make accurate local temperature measurements. In addition, high electric powers are involved, usually associated with high voltages and/or high currents, which is a challenge in choosing an appropriate thermometry. Various temperature measurement methods have been explored for this application, including doped single crystal silicon resistive thermometers, isolated poly silicon resistive thermometers, and infrared thermometry. The advantages and disadvantages of each method are discussed in this section. From experimental experiences, using a single resistor strip as both a heater and a series of thermometers, or aluminum as heaters and doped single crystal silicon resistors as thermometers is the best for high-power heat transfer measurements, as described in Sects. 3.3.1 and 3.3.2.

3.3.1 Doped Silicon Resistor as Both Heater and Thermometers

In this method the temperature is measured from the temperature coefficients of doped single crystal silicon resistors. As shown in Fig. 3.6, a single resistor is formed by ion implantation, and divided into a few segments with aluminum

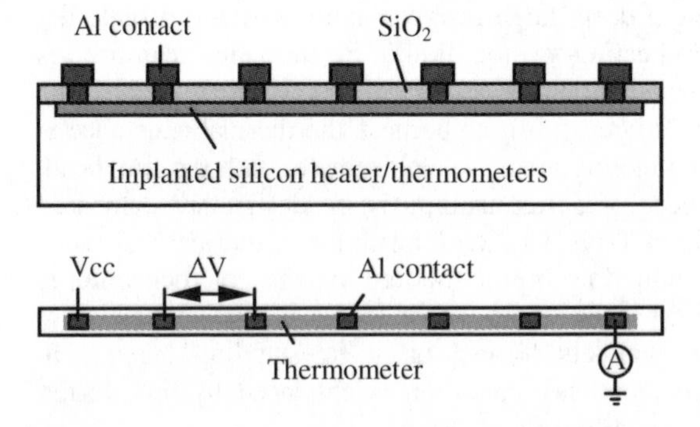

Fig. 3.6 Schematic of a single resistor functioning as both a heater and a series of thermometers. The resistor is formed by heavily doped single crystal silicon.

contacts. In application, the entire resistor is applied a constant current by the power supply *Vcc*, and the current is measured with a current meter. The voltage difference ΔV across each thermometer (or a segment of the heater) can be recorded in heat transfer experiments. The voltage difference is then divided by the current to obtain the individual thermometer resistance. By comparing the resistance change with pre-calibrated temperature response of the thermometer, the temperature information can be obtained.

As shown in Fig. 3.7, doped single crystal silicon resistors are very good temperature sensors because of their temperature dependent resistance. The mobility of carriers (electrons and holes) in semiconductors is a function of temperature. Depending on the type and concentration of the dopant, which provides electrons or holes to the semiconductor, the resistance of the semiconductor varies as a function of the temperature with a linear or polynomial relationship. This characteristic correlation remains once established when the resistor is formed, and will not be affected by the environment.

Boron doped P-type resistors on an N-type substrate are more often used because they tend to form ohmic contacts to aluminum. However, there is no reason why N-type resistors cannot be used. N-type resistors may have even higher temperature sensitivities, but they often require additional doping at the contact region for a good ohmic contact, which complicates the fabrication process.

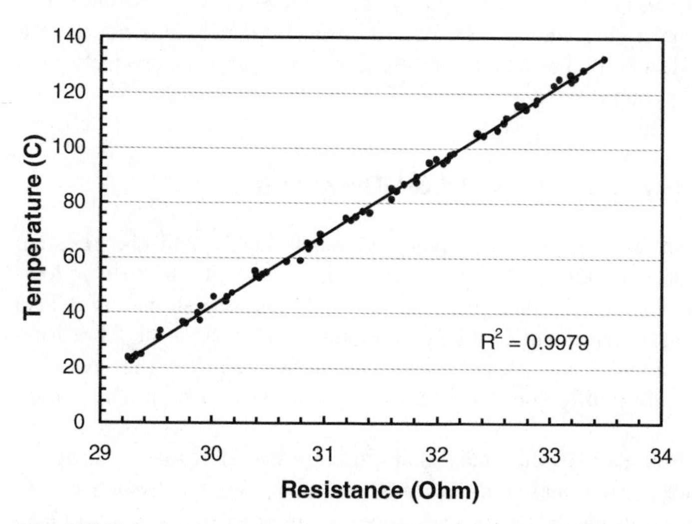

Fig. 3.7 Temperature calibration and the fitting curve of a heavily doped N-type resistive thermometer. Figure shows data recorded during six independent calibrations of the same resistor, illustrating repeatability and linearity. The thermometer has 0.04 Ω/°C sensitivity and ± 3 °C calibration accuracy.

Compactness is the main advantage of this design. The integration of the heater and the thermometers minimizes the required space. Because only one resistor strip is required, the freestanding beam can be made very narrow, resulting in better thermal isolation and constant heat flux boundary conditions.

On the other hand, there is a trade-off between the temperature sensitivity and the functionality of the device. The conflict lies in the contradictory requirements of the heater and the thermometers—the heater resistance requires to be low to avoid device damage at high power inputs due to the PN junction breakdown; while the thermometers need high resistance for high temperature sensitivity. However, the resistance of the thermometers is only a fraction of the heater resistance from the design. As a result, the temperature sensitivity can be very low due to the limited heater resistance. As the calibration curve shows in Fig. 3.7, with a total resistance of about 30 Ω, the thermometer only shows 0.04 Ω/°C temperature sensitivity.

And it is difficult to implement automatic data acquisition to this design, because of the high voltage involvement and low temperature sensitivity. Even at very high doping concentration, for example, 10 Ω/□ sheet resistance, the bulk resistance of a 2 cm long, 400 μm wide resistor reaches 500 Ω. Considering that a normal A/D card allows ±10 V voltage input, the maximum power dissipation from the resistor at 10 V is only 0.2 W, while the heat transfer experiment with the microchannels may require as much as 10 W heat rate. Efforts have been made to find a proper yet not very expensive way to accurately measure the voltage, including voltage dividers with instrumentation Op-amps, which will be discussed in detail later. Unfortunately, a tiny drift of the voltage divider resistance can easily bury the temperature signals. However, automatic data acquisition is still possible for this design if the signal conditioning circuit is very precisely (and maybe expensively) compensated.

3.3.2 Aluminum Heaters and Doped Silicon Thermometers

This design shown in Fig. 3.8 decouples the metal heater and the resistive thermometers; hence provides very high temperature sensitivity as well as low-voltage heating, ready for automatic data acquisition. Because aluminum is also used as contacts, it is convenient to integrate aluminum heaters in the fabrication process without adding an extra photolithography step. The resistive thermometers are only lightly doped to yield high resistance, or high temperature sensitivity.

The thermometers also require individual calibrations before use. However, automatic data acquisition makes the calibration much more convenient. As shown in Fig. 3.9, with nearly 10 kΩ resistance, the thermometer has 5-10 Ω/°C temperature sensitivity over a 25-140 °C temperature range. Although the temperature characteristic curve is no longer linear compared with the heavily doped thermometer shown in Fig. 3.8, the calculated fitting curve has ±0.5 °C accuracy, which is reasonable for this application.

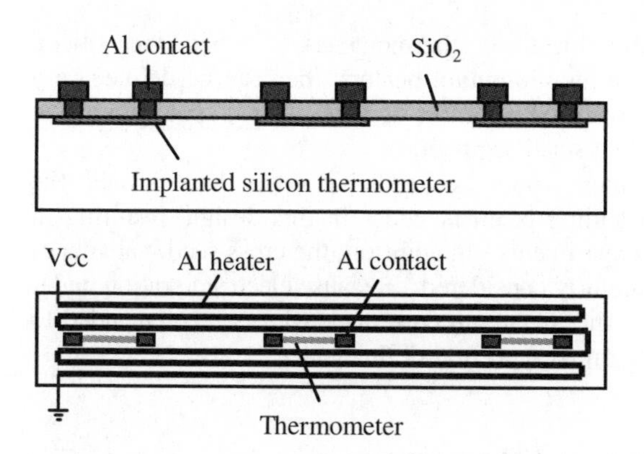

Fig. 3.8 Schematic of the aluminum heaters and silicon thermometers. The thermometers are implanted first, and aluminum is deposited on a SiO_2 insulation layer above the thermometers, forming both contacts and the heater.

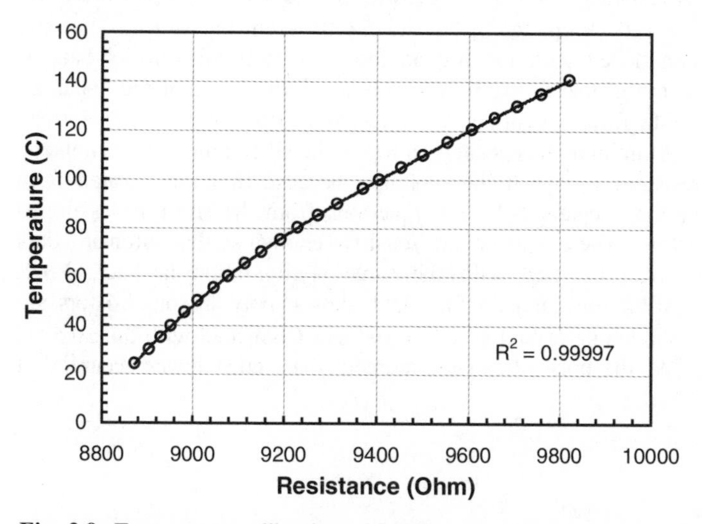

Fig. 3.9 Temperature calibration with data acquisition system and the fitting curve of a lightly doped N-type resistive thermometer. The temperature sensitivity varies from 5 Ω/°C to 10 Ω/°C in the full range, with ± 0.5 °C calibration accuracy.

This configuration is well designed for transient measurements that require automatic data acquisition. It is primarily used in experiments presented in Chaps. 5 and 6. Because the high-resistance thermometers yield very good temperature sensitivity (5-10 Ω/°C, shown in Fig. 3.9) as well as low noise, this design is highly recommended for similar applications.

Another advantage is more localized thermometers. Since the silicon thermometers are separated from the aluminum heaters, they can be defined very small in size to measure local wall temperatures at a very small spot rather than spatially averaged temperature in a small segment.

The disadvantage is that more room is needed for the heaters and the thermometers; hence the freestanding beam is wider in this design, resulting in larger amounts of heat loss in experiments. In addition, the cross sectional area of the aluminum wires must be carefully considered. Because electromigration under high current can melt the wire, the maximum current density is recommended to be smaller than 10^{10} A/m^2 at the thinnest section of the heater.

3.3.3 Doped Poly Silicon Heaters and Thermometers

Doped poly silicon resistors can be used to replace the doped single crystal silicon resistors in both designs described in Sects. 3.3.1 and 3.3.2. As shown in Fig. 3.10, a 0.5-1 μm thick poly silicon film is deposited on a silicon nitride insulation layer as resistors. Nitride is preferred because it has higher dielectric strength than oxide (10^7 V/cm for Si_3N_4 and 5×10^6 V/cm for SiO_2). The poly silicon film is then doped with ion implantation. A poly silicon etch-back is followed to define the geometry, so the resistors are formed as isolated "islands" on the silicon nitride film. A silicon dioxide passivation layer is then formed during annealing. Aluminum is deposited on top of the silicon dioxide as contacts.

Poly silicon resistors survive higher voltages because they are isolated from each other, rather than separated by PN junctions from the substrate as doped single crystal resistors. These resistors are sensitive enough so that automatic data acquisition system can be used with the same configuration in Sect. 3.3.2. However, as the calibration curve in Fig. 3.11 shows, poly silicon resistors are generally noisier than single crystal silicon resistors. Compared with the ±0.5 °C accuracy in Fig. 3.9, the poly silicon thermometer has much lower accuracy of ±2 °C.

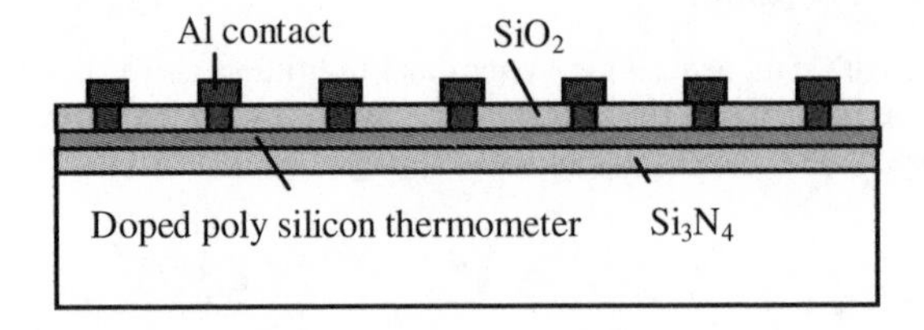

Fig. 3.10 Schematic of doped poly silicon thermometers. The resistive thermometers are formed on a silicon nitride insulation layer and doped with ion implantation. Silicon dioxide is formed during annealing and aluminum is then deposited on top of the silicon dioxide film.

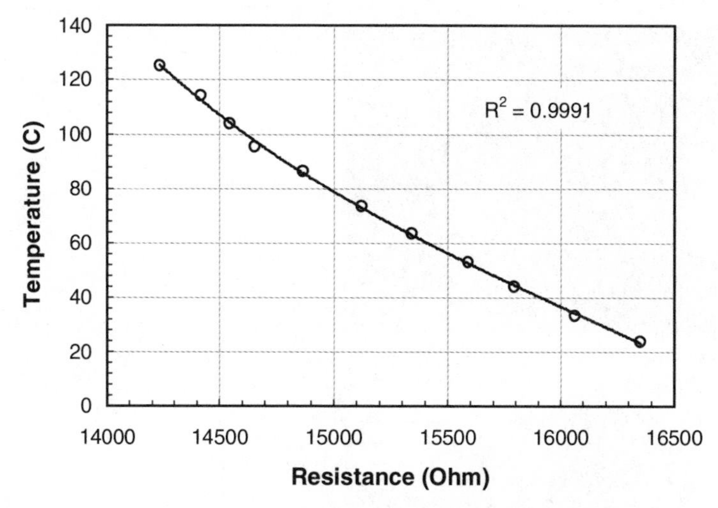

Fig. 3.11 Temperature calibration and fitting curve of a lightly doped N-type poly silicon thermometer. Unlike the single crystal silicon, boron doped poly silicon shows an inverse relation to the temperature. This thermometer has a temperature sensitivity of –10 to –30 Ω/°C, with ±2 °C calibration accuracy.

3.3.4 Other Thermometry

There are many commercially available thermometers or temperature measurement equipment, for example, thermocouples, thermisters, and infrared cameras, which have been used in various thermal experiments. For example, Peng et al. [2.9] used thermocouple to measure the wall temperature distribution in the study of two-phase flow in small channels made of stainless steel. The experimental data were noisy, possibly due to the contact resistance between the thermocouple and the steel surface. Because most commercial temperature sensors require external attachment to the sample by thermal grease or epoxy, the accuracy is not comparable with the silicon thermometers that are integrated in the bottom wall of the test channels.

One promising method is infrared thermometry, which is capable of providing the entire temperature field of a silicon chip. An infrared camera was used in the initial experiments with the single-phase heat sink chips described in Chap. 2. Figure 3.12 is the temperature field of a heat sink chip at 25 W input heat rate with 5 ml/min water flow rate, showing the wall temperature increase from the

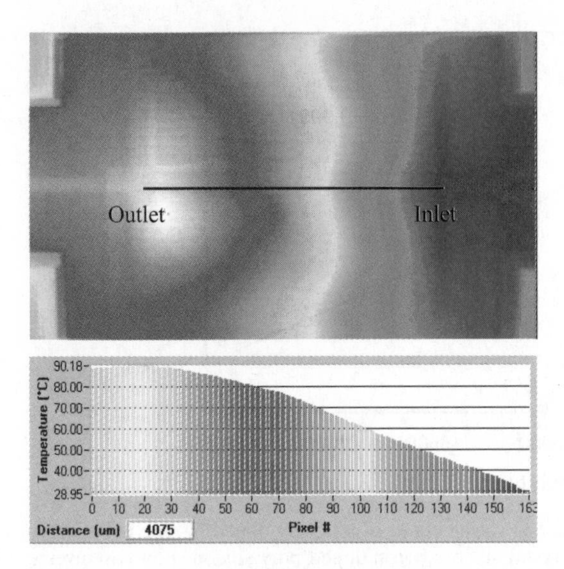

Fig. 3.12 Infrared images of the temperature field of a heat sink chip with 25 W input heat rate and 5 ml/min DI water flow rate. The top image is the plan view of the temperature field, and the bottom image is the temperature profile on the line drawn in the plan view.

inlet to the outlet direction. However, a major difficulty with infrared thermometry is the temperature calibration of the samples. The standard calibration requires that the sample be placed facing up on a constant temperature heater provided with the camera, which is not applicable to the instrumented microchannel chip because of the wire bonds on the back side. In the measurement in Fig. 3.12, hot water was pumped into the channels and the thermometers on the chip were used as temperature references. Since the sample was not ideally kept isothermal, measurement errors were unavoidable. On the other hand, because the bonded silicon-glass chip is very thin, only about 1 mm in thickness, the sample is translucent under the infrared camera, which significantly increases the background noise under high magnifications. Despite these problems, infrared thermometry will be the first choice if the measurement of the entire temperature field is necessary.

3.4 Experimental and Data Acquisition Systems

The experimental system consists of a fixture to support and direct liquid into the single-channel device, temperature and pressure sensing elements, optical equipment for visualizations, and an automatic data acquisition system for both steady and transient state measurements.

3.4.1 Experimental System Configuration

As shown in Fig. 3.13, the device is mounted on the fixture and placed under a microscope with a CCD camera. The fixture is made of Ultem, a material with low thermal conductivity (k=0.22 W/m-K), with lower heat loss from the test channel. The outlet of the fixture is open to the air and a 100 μm diameter thermocouple is inserted to measure exit water temperature. A MSP400 pressure transducer is used to measure the pressure drop along the test channel. DI water is supplied to the system by a Harvard PHD2000 syringe pump. Pressure, exit fluid temperature, and the wall temperatures provided by the integrated thermometers are sampled with a National Instruments DAQCard-AI-16XE-50 A/D card and analyzed in LabVIEW.

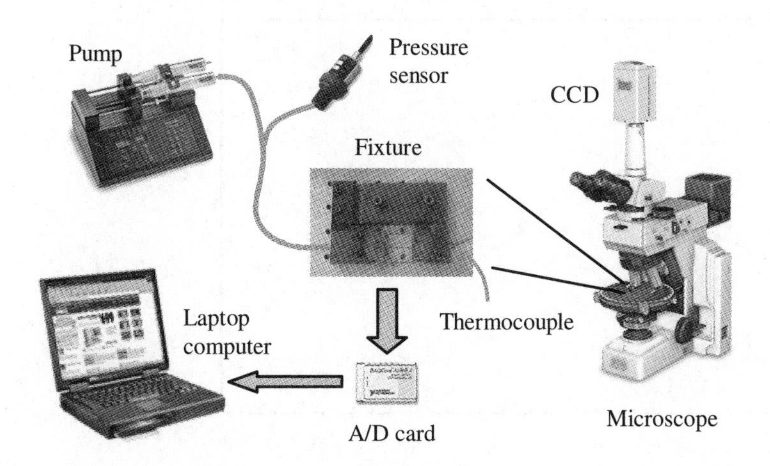

Fig. 3.13 Diagram of the experimental system. The instrumented channel chip is mounted on a low thermal conductivity fixture and placed under the microscope with a CCD camera. The pressure and temperature signals are collected with an A/D data acquisition card and processed with a laptop computer.

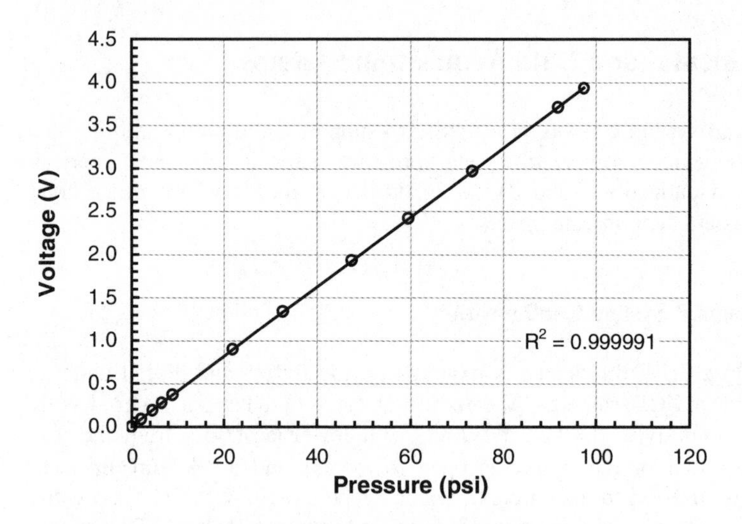

Fig. 3.14 An MSP400 pressure transducer calibration and fitting curve. It is calibrated with a pressure calibration unit which supplies a standard pressure from 0-100 psi, and the output of the transducer is recorded. The output gives ±0.1 psi measurement error.

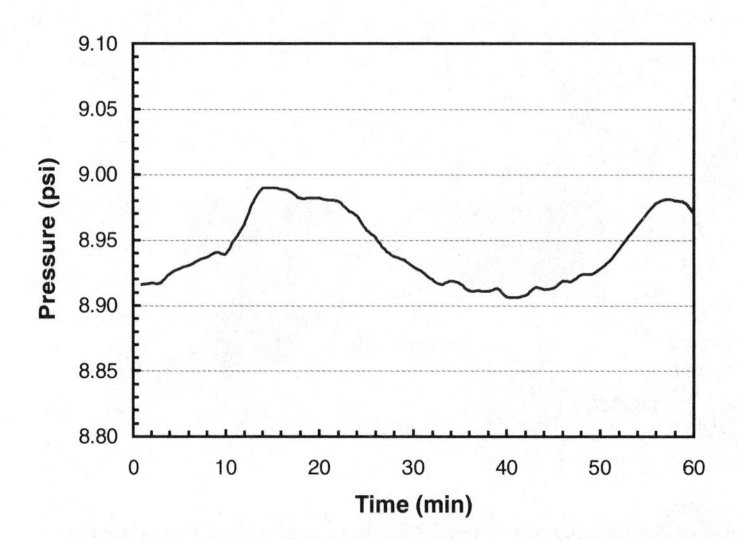

Fig. 3.15 Long-term stability of the syringe pump. The pump was programmed to supply a constant flow rate of 0.1 ml/min to a microchannel, and the pressure drop in the channel was recorded every minute for a one-hour period. The ±0.5% pressure fluctuations indicate same percentage of flow rate fluctuations.

Pressure Transducer

The pressure transducer MSP400-100-P-3 is a micromachined sensor by Measurement Specialties Inc. The measurement range is 0-100 psi gauge pressure. Depending on the supply voltage (5-15 V), the transducer provides 0.5-4.5 V full-scale outputs. Figure 3.14 shows the sensor calibration with a pressure calibration unit at 12 V voltage supply. As shown in Fig. 3.14, the pressure transducer has a very good linearity of 0.1%, or ±0.1 psi absolute error, over the entire measurement range.

Syringe Pump

A Harvard PHD2000 syringe pump was selected to provide a constant DI water flow to the system. Since a constant flow rate is critical to the two-phase flow in microchannels, the pump flow rate with pressure load must be examined before use. As shown in Fig. 3.15, the pump has about ±0.5% flow rate fluctuations at the steady state with a constant pressure load.

CCD Camera

Because of the very small channel areas, the bubble nucleation process happens in a very short time scale and the vapor flow can move as fast as 100 m/s in less than 100 μm diameter channels. Continuous CCD cameras without a shutter cannot clearly capture transient events on this scale because the maximum object velocity v that is allowed for a clear image is determined by the point response function d_{prf} and the light integration time t by $v=d_{prf}/t$, and $d_{prf}=1.22\lambda/NA$, where λ is the wavelength of the light (assumed to be green light with λ=510 nm) and NA is the numerical aperture of the microscope objective lens. As an example, with a 10X magnification and NA=0.25 objective lens, a continuous CCD camera with 30 frames/s frame rate, or integration time t=0.033s, the maximum allowed object speed is 75 μm/s. However, with a shutter controlled CCD camera that allows 100 frames/s, or t=0.01s, even objects moving at 250 μm/s can be clearly imaged. A SONY 2735 CCD camera with maximum t=1/100000s shutter speed is used for flow regime visualizations discussed in Chap. 5.

3.4.2 Data Acquisition System Configuration

The data acquisition system consists of a laptop computer and an A/D card. Considering the data accuracy and number of analog input channels, a 16-bit, 16-channel PCMCIA DAQCard-AI-16XE-50 is a good choice. At 20 kS/s maximum sampling rate, the A/D card is capable of continuously scanning all 16 channels at 1000 Hz frequency. The card can be configured as 16 single-ended or 8 differential input channels. The resolution of A/D conversion depends on the input range of the channel. At 0-10 V input range, the resolution of the A/D conversion is $10/2^{16}$=0.15 (mV).

Table 3.2 DAQ Card pin configurations.

Signal	Channel configuration
Pressure (pressure transducer)	Differential input
Exit water temperature (thermocouple)	Differential input
Bias voltage for thermometers	Single-ended non-referenced
Temperature distribution R1 through R7	Single-ended non-referenced
Heater current	Single-ended ground referenced
Heater voltage	Single-ended ground referenced

For the heater/thermometer design in Sect. 3.3.2, there are a total of seven thermometers and one heater on the chip. In addition, a voltage output pressure transducer and a K-type voltage output thermocouple (already converted to Celsius degree by a thermocouple module) also requires sampling in the experimental system. The A/D card channel configurations for each signal are given in Table 3.2, and the schematic is shown in Fig. 3.16.

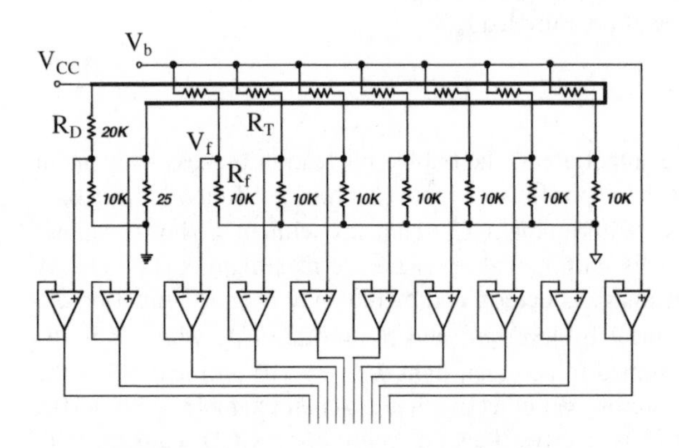

Fig. 3.16 Schematic of signals to A/D card connections. The heater and the thermometers have separated grounds to eliminate noise. The heater is connected to the real ground, and the thermometers are electronically grounded.

The pressure transducer and the thermocouple outputs are directly connected to two differential channels. As shown in Fig. 3.16, all thermometers share a common bias voltage that provides the bias current to the resistors. A reference resistor R_f with known resistance is connected to each of the thermometers. After V_f and V_b are sampled, the bias current is calculated from $I_b=V_f/R_f$, and the thermometer resistance can be found from $R_T=(V_b-V_f)/I_b$. The heater power is measured by the same means, with a reference power resistor to measure the current. A voltage divider is used to protect the A/D card when the heater input voltage exceeds 10 V.

This configuration is recommended for the thermometry design in Sect. 3.3.2. All the transient measurements discussed later in the book rely on this data acquisition system. However, careless configurations of the A/D channels and the signal conditioning circuit can significantly affect the measurement accuracy. The following sections discuss related issues and how they affect the measurement results.

Thermometer Resistance Measurement

Figure 3.17 is typical temperature calibration curves on a single-channel device—there are some resistance variations after ion implantation, but all seven thermometers have similar temperature dependence. The measurement was made with the standard configuration shown in Fig. 3.16. An alternative approach has been explored, called "series configuration" because the thermometers share the same bias current. The schematic is shown in Fig. 3.18.

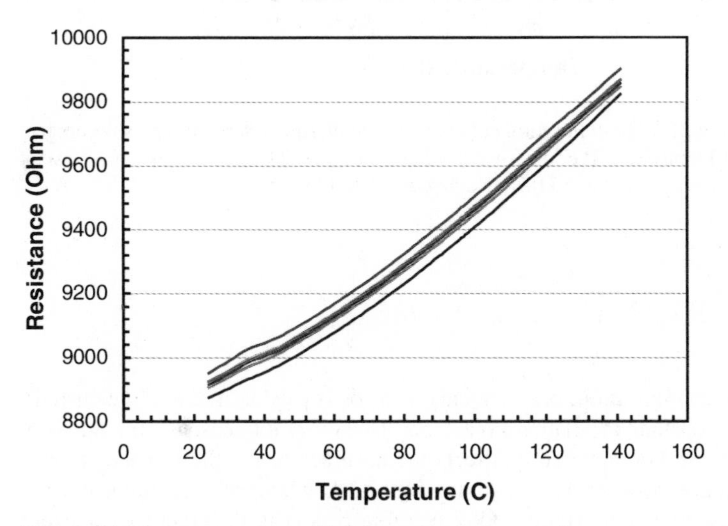

Fig. 3.17 The temperature calibration curves of seven thermometers on the same device, with parallel configuration as in Fig. 3.16.

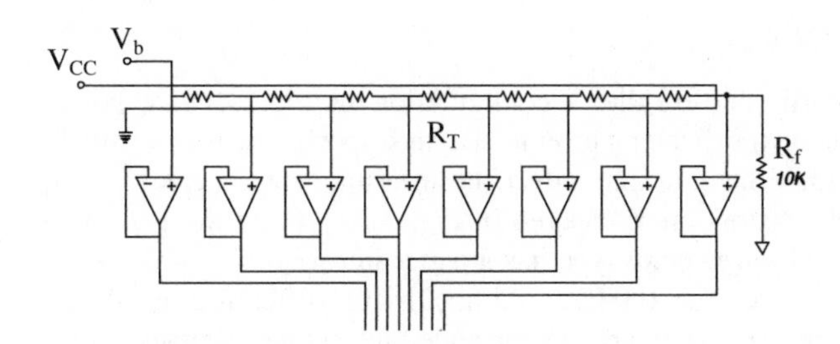

Fig. 3.18 Series configuration of the thermometers. All thermometers have the same bias current, and one reference resistor R_f is used to measure the current.

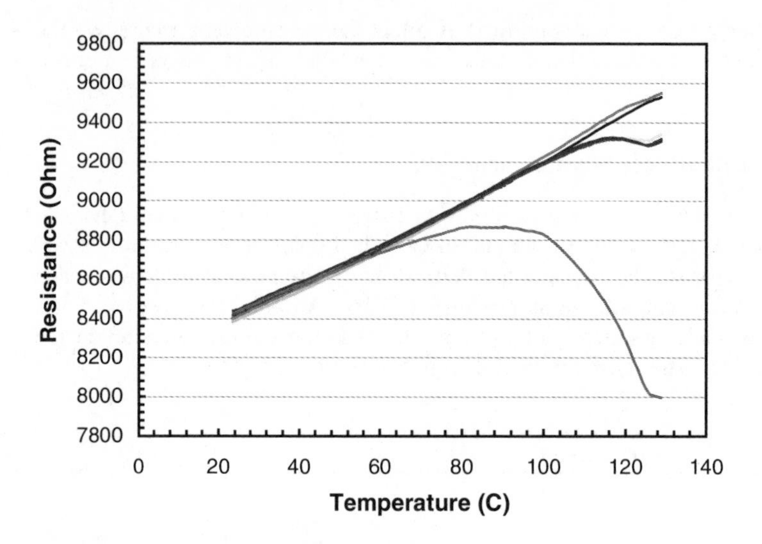

Fig. 3.19 The temperature calibration curves of seven thermometers on the same device, with series configuration. The measurement error is possibly caused by non-constant current in the thermometers due to the leakage current to Op-amps.

In the series configuration, one reference resistor is used to measure the current, assumed to be constant in all thermometers. However, for reasons still not very clear, this configuration apparently affects the accuracy of the thermometers away from the reference resistor as shown in Fig. 3.19, where all thermometers are expected to have the same trend. One possible reason is that there are current losses to the Op-amps in each measurement point, causing a non-constant current in the thermometers.

Heater Power Measurement

As shown in the standard (or "parallel") configuration in Fig. 3.16, a power resistor is used to measure the heater current. Because power resistors usually have small resistances, joule heating may affect the measurement accuracy. Two types of common power resistors are compared in Fig. 3.20. From the I-V curve fittings, the silicone coated wirewound type has a better linearity of 0.04%, while the ceramic composition type has 0.08%. Although the difference is almost negligible, wirewound power resistors are recommended for higher accuracy.

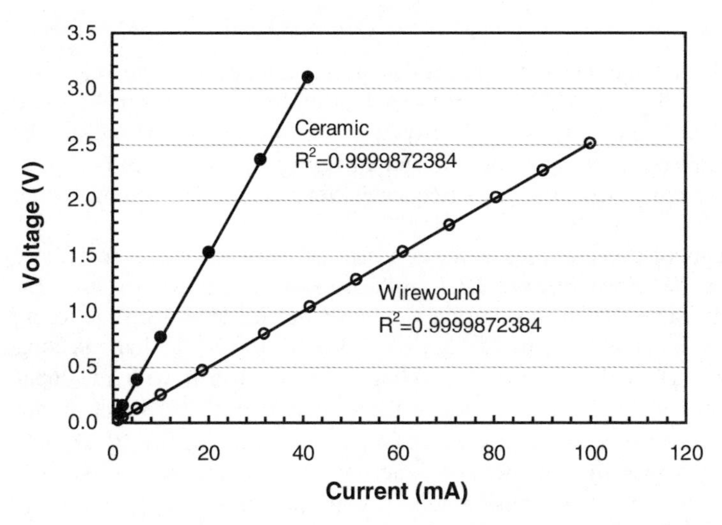

Fig. 3.20 I-V curves of a silicone coated wirewound power resistor and a ceramic composition power resistor. The wirewound type has better linearity of 0.04% compared with 0.08% of the ceramic type.

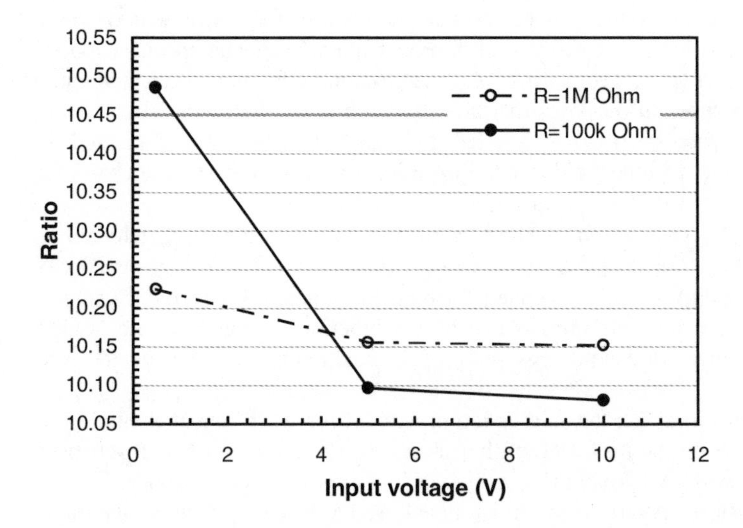

Fig. 3.21 The ratio of voltage dividers as a function of input voltage. The change of the ratio results from the Joule heating in the resistors. The accuracy of the voltage conversion is good for a certain range of input voltages, but inconsistency is significant over large scales of input voltage.

Discussions

To lower the signal white noise during the data acquisition, an averaging algorism can be implemented in LabVIEW. 1000-sample average at 1 kHz sampling rate per channel yields good results for steady state measurements. However, for transient measurements, the sampling rate and averaging must be carefully chosen because the averaging may lose high-frequency components that belong to the signal.

In the attempts of implementing automatic data acquisition for low sensitivity thermometers as described in Sect 3.3.1, voltage dividers like the one used for heater power measurement in this section were considered. However, due to the large span of the input voltage, which may range from 5-100 V, the joule heating in the resistors causes the ratio of the voltage divider to drift with the input voltage, as shown in Fig. 3.21. Although the drift is only 0.2%, which is not significant for the heater power measurement, it results in up to 10 Ω drift equivalent in resistance measurement, making the temperature signal very noisy and inaccurate due to the low temperature sensitivity (less than 0.05 Ω/°C) of the thermometers.

3.5 Fabrication of Instrumented Single-channel Devices

This section introduces the general fabrication process for instrumented single channel test chips, including discussion of process parameters and special methods used in making irregular geometries. All the structures that have been fabricated for various experimental purposes that are discussed in the following chapters are based on this process flow. Detailed process parameters are discussed in individual chapters together with related experiments. The process flow charts for specific devices are listed in the appendices.

N type, <100> orientation, 400-500 μm thick silicon wafers are selected as device substrates. For doped resistive heaters (as in Sect. 3.3.1), high resistivity wafers (>100 Ω-cm) are recommended for high PN junction breakdown voltages. Because the doped resistors are isolated by PN junctions in the substrate, a high breakdown voltage threshold ensure proper functioning of the heater and thermometers.

As shown in Fig. 3.22, the fabrication process begins with ion implantation to form the resistors on the back side of the substrate. The dopants are then activated and driven-in with wet oxidation, which also forms a silicon dioxide film to protect the resistors. Again, if the resistors are used as heaters, a long annealing is required to deepen the doping profile for higher breakdown voltage thresholds.

Next, contact windows are opened in the oxide layer and 99% aluminum with 1% Si is deposited as contacts. Since boron oxide may form during the high-dose annealing process, an argon etch prior to deposition is necessary in order to form a good contact. After aluminum etch, a freckle etch is performed to remove residual aluminum traces from the substrate surface to prevent shorts between contacts.

The final ohmic contact is formed by a 1-hour forming gas annealing at 400 °C.

To protect aluminum heaters and contact pads, 2500 Å LTO (Low Temperature Oxide) is deposited and then etched-back, to expose the wire bonding pads. The LTO film also helps to protect aluminum heaters from electromigration damages.

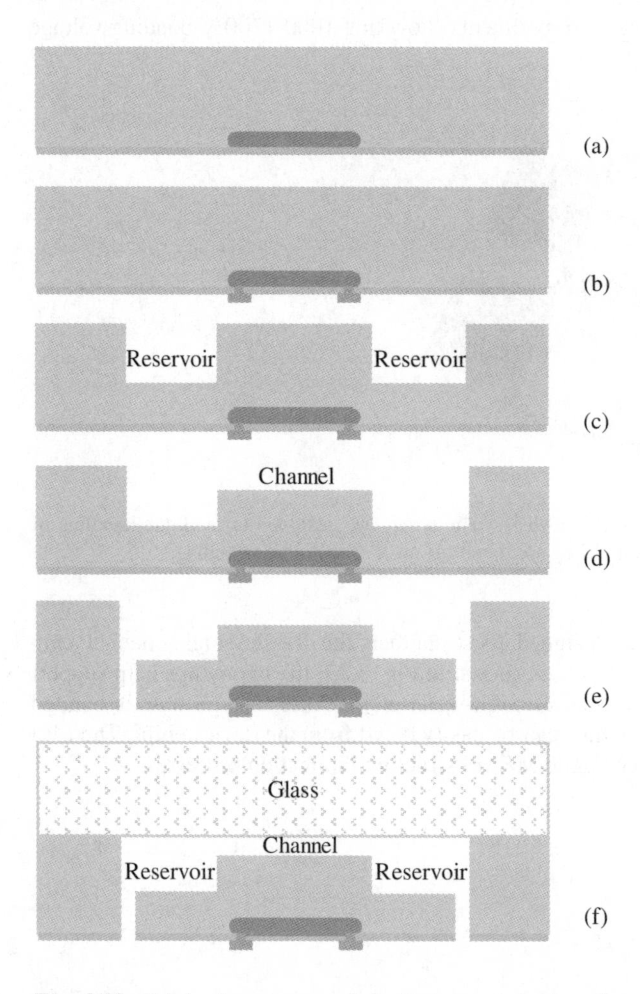

Fig. 3.22 Fabrication process of the instrumented microchannels. (a) Ion implantation. (b) Annealing and Al deposition. (c) Reservoir etch with DRIE. (d) Channel etch. (e) Back side etch to open the fluid inlet and outlet. (f) Silicon-glass anodic bonding.

The process is then continued with etching the reservoirs and channels from the front side. The deep reservoirs are first DRIE (Deep Reactive Ion Etching) etched into the substrate. Then 7 μm thick photoresist (SPR220-7 by Shipley Inc.) is used as the channel etching mask, which is thick enough to cover the 250 μm deep trenches. The depth of the channel is controlled by a timed etch and inspection. Next, the wafer is temporarily bonded to a support wafer with photoresist for the back side through-etch to create the inlet/outlet holes, as well as to separate the device chips from the substrate.

Finally, the device chip is anodically bonded to a pre-diced Pyrex 7740 piece at 250 °C and 1200 V voltage. Experiments show that 1000-1200 V bonding voltage is safe for the resistors.

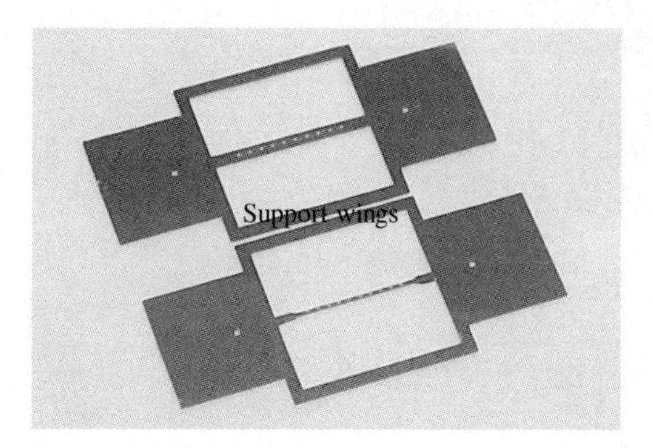

Fig. 3.23 Support wings on the devices. The wings are designed to facilitate handling of the device during fabrication. They are diced off after silicon-glass bonding.

A winged structure is designed to strengthen the fragile single-channel chip during the fabrication process. As shown in Fig. 3.23, the two wings help support the device chips after the final through-wafer etch. After the substrate is bonded to glass, the 2 mm wide wings can be easily diced from the device chip. The idea can be used in many other fragile MEMS structure fabrication processes.

4 Measurements and Modeling of Two-phase Flow in Microchannels

4.1 Review of Previous Research

While the modeling and optimization of single-phase liquid microchannel heat sinks have been extensively studied [2.11-2.14, 4.1-4.3], the more complicated physics of boiling flow in microchannels is relatively poorly understood. Unlike in the single-phase forced convective flow, the two-phase convective coefficient is no longer a constant for fully developed conditions, but closely coupled with local flow conditions. Depending on how the vapor and liquid mixture moves in the channel, two basic flow models have been used for pressure drop simulations. The homogeneous model assumes that the liquid and vapor move at the same velocity at every position along the flow direction. The other approach, the annular model, assumes that a thin, slow-moving liquid film surrounds a rapidly moving vapor core. For saturated boiling heat transfer in macroscale horizontal tubes, the most widely accepted heat transfer correlation is Kandlikar's correlation, which has relatively good agreement with data for a broad spectrum of fluids over a wide range of conditions [4.4]. Other correlations include the Shah correlation, the Chen correlation, the Schrock and Grossman correlation, the Gungor and Winterton correlation, and the Bjorge, Hall and Rohsenow correlation, each of which is based upon certain two-phase heat transfer data [4.4]. Since all flow models and heat transfer correlations are determined by the flow conditions, it is important to understand the microchannel flow behavior as well as to obtain reliable heat transfer data as a reference to select the appropriate model.

Lee et al. studied water and air mixture flow at room temperature in horizontal channels between two plates, with 200 μm-5 mm hydraulic diameters [4.5]. They proposed a Lockhart-Martinelli type pressure correlation for water and gas two-phase flow without heat transfer or phase change. The model has a modified correction factor for microchannels of the studied range.

Stanley et al. conducted two-phase flow experiments in rectangular aluminum channels with 56-256 μm diameters to study the pressure drop and the heat transfer [4.6]. Their experiments did not involve the phase change, but focused on the heat transfer properties of two-phase flows. Inert gases were mixed with water to eliminate evaporation and minimize variations in the mass flow rates of the liquid and gas phases. The measured pressure drop and heat transfer data were consistent with a homogeneous flow model, in which case the liquid and gas flow velocities are considered identical.

Riehl et al. reviewed and compared heat transfer correlations for both single- and two-phase flows with Reynolds numbers ranging from 300 to 80,000 [4.7]. They simulated the Nusselt number from several correlations under three arbitrary mixture qualities. The results show that for low mixture qualities, the correlations provide similar results, but further study is necessary for high-quality mixtures.

Peles et al. developed a one-dimensional flow model based on their experiments with 50-200 μm hydraulic diameter triangular channels made in silicon substrates [4.8]. From the observation of a flat evaporation interface dividing the liquid and vapor into two distinct domains, their model treats the pure liquid phase and the pure vapor phase as separate flows.

Jacobi et al. proposed a heat transfer model assuming the primary heat transfer mechanism is thin-film evaporation into elongated bubbles, and validated this assumption from two-phase heat transfer measurements in 1.95 mm and 2.46 mm diameter tubes with several refrigerants as the cooling liquid [4.9].

Some other two-phase heat transfer models have also been studied in microchannels with larger hydraulic diameters [4.10-4.12]. However, a more general model that agrees with experimental data in microchannels has not been proposed. This chapter discusses the wall temperature measurement made on 28-60 μm diameter channels with nearly-constant heat flux boundary conditions and a one-dimensional finite volume model for single- and two-phase heat transfer.

4.2 Design Parameters of Test Devices

Figure 4.1 shows the schematic of the test device based on the conceptual design proposed in Chap. 3. The overall chip dimension is 2×6.5×0.05 cm. The outlet of the microchannel is formed in a 3×3 mm pad at the end of the channel so that the exit water temperature is directly measured with a thermocouple before there is significant heat loss to the surrounding silicon substrate. In order to study phase change in microchannels and boiling uniformity across a group of channels, both single-channel and multi-channel devices were fabricated. One resistor as both a heater and series of thermometers as described in Sect. 3.3.1 is used in the

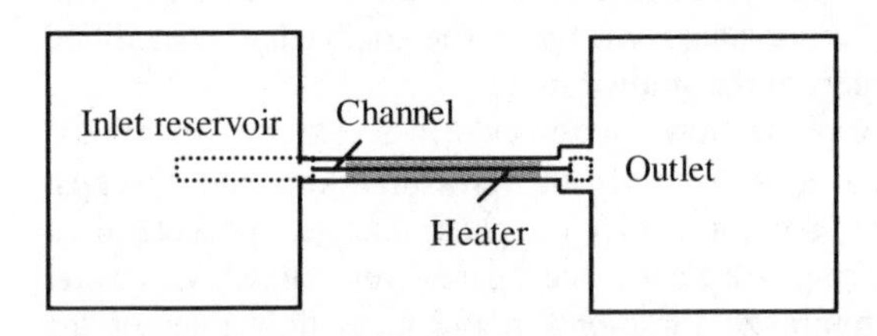

Fig. 4.1 Schematic of the test device, showing the arrangement of the inlet reservoir and outlet port and the freestanding, instrumented microchannel.

design to minimize the width of the freestanding beam, hence to maintain nearly-constant heat flux boundary conditions. Design parameters are listed in Table 4.1.

Table 4.1 Structural parameters of freestanding microchannels with a single resistor as both a heater and thermometers.

Structure	Dimension
Inlet reservoir	600 μm wide, 250 μm deep, 1 cm long
Freestanding beam	500 μm (single) / 2 mm (multi) wide, 2 cm long
Single-channel design	40-50 μm wide, 50-100 μm deep, 2 cm long
Multi-channel design	20 μm (width) x 50-70 μm (depth) x 40, 40 μm (width) x 100 μm (depth) x 20, 2 cm long
Resistor (heater)	400 μm wide, 1.6 cm long (single) 1 mm wide, 1.8 cm long (multi) 800 Ω (single) and 360 Ω (multi) resistor depth 7 μm
Si substrate thickness	400 μm
Glass thickness	520 μm

Figure 4.2 shows the cross sectional view of the device. In order to obtain a high PN junction breakdown voltage threshold, 1000 Ω-cm resistivity N-type substrates were used and the doped resistor was annealed for 6 hours at 1150 °C after ion implantation. The simulated doping profile is 7 μm deep and more than 100 V breakdown voltage is expected. The process flow chart is listed in Appendix A.

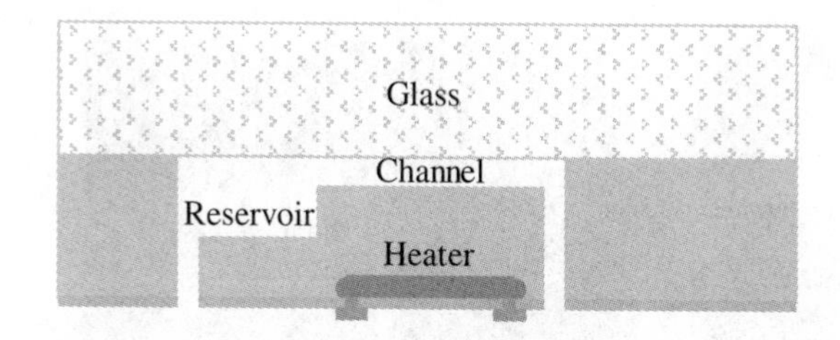

Fig. 4.2 Cross sectional view of the test channel with a resistor as both a heater and nine thermometers. There are actually 10 aluminum contacts, of which only two are shown in the schematic.

More detailed views of the channel region are provided in Fig. 4.3. A 600 μm wide, 250 μm deep reservoir is defined at the inlet, with a 2 mm long entrance region (Fig. 4.3a). The entrance region is the part of the microchannel that is not heated, which helps to reduce the pre-heating of the fluid. There is a similar exit region at the outlet. A single resistor (Fig. 4.3b) is formed on the back side of the channel. The resistor, used as the heater, is divided into nine segments, each of which functions as a thermometer.

Figure 4.4 shows some SEM images of the single- and multi-channel designs. The channels are rectangular in shape with plain side walls. Some over-etch in the bottom of the reservoir is shown, which is caused by two DRIE etch steps. The over-etch does not affect the device functionality and may be avoided by careful photoresist coating, or selection of a photoresist with higher viscosity.

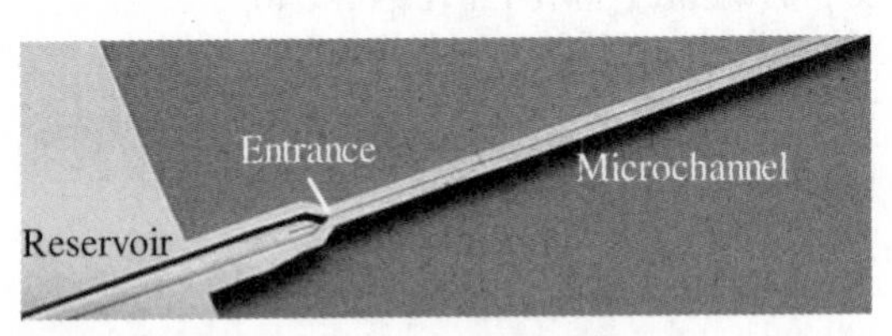

(a) Front side of a single-channel device.

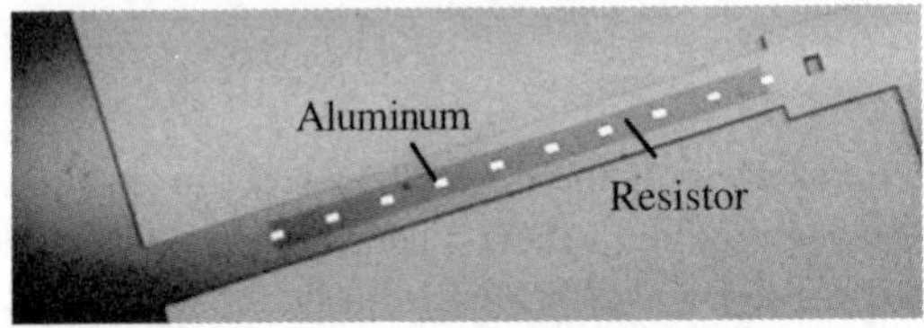

(b) The heater/thermometers on the back side of a multi-channel device.

Fig. 4.3 Close-up views of the channel geometry and the doped resistor.

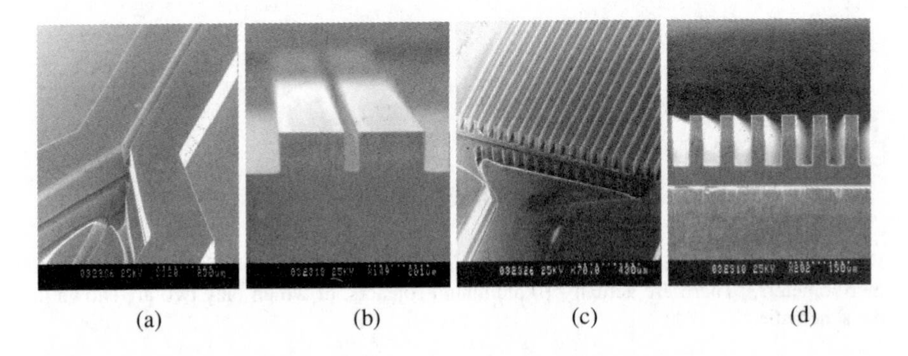

Fig. 4.4 SEM images of the single-channel and multi-channel devices. (a)-(b) The entrance region and cross section of a 40 μm wide, 100 μm deep channel. (c)-(d) The inlet reservoir and cross section of a 40-channels device, each channel is 40 μm wide and 100 μm deep.

4.3 Two-phase Flow Measurements

Various devices with 28-60 μm diameter channels have been tested under constant water flow rates. These experiments were conducted to obtain accurate pressure and wall temperature distribution measurements during the phase change, as supporting data for the flow and heat transfer models.

Measurement results of two devices are presented in this section. Device No. 1 is a 40-channel design, with individual channel dimensions of 20 μm wide and 70 μm deep (31 μm hydraulic diameter). Device No. 2 is a 50 μm wide and 70 μm deep (58 μm hydraulic diameter) single-channel design. Both results were measured at 0.1 ml/min constant DI wafer flow rate.

4.3.1 Measurement Error Analysis

As discussed in Chap. 3, automatic data acquisition is not applicable for the single resistor design. The measurements were performed manually with multimeters at steady states. As shown in the schematic in Sect. 3.3.1, a power supply Vcc is used to supply a constant current to the heater, and the current is measured with a current meter. The voltage difference ΔV across each thermometer is measured with a voltage meter. The resistances of the thermometers are then taken to the calibrated resistance-temperature function to calculate the corresponding temperatures. The temperature calibration curve in Fig. 3.7 is obtained from a typical multi-channel device, showing 0.04 Ω/°C temperature sensitivity. Thermometers on the single channel devices have the same characteristics, with larger resistances (70-80 Ω) and higher sensitivity of 0.1 Ω/°C.

The main error source of the wall temperature measurement comes from the uncertainties in the temperature calibration and curve fitting, which is ±3 °C from the mean square deviation. The absolute value is larger then that of the thermometers in the heat sink chips (±2 °C); however, because of the improved structure design, the experimental conditions are much better controlled, and the wall temperature profile measurement has a higher resolution.

4.3.2 Multi-channel Device Measurements

The multi-channel devices are designed to ensure the boiling uniformity across the channels because the channels are closer to each other and the heat flux is more uniform. Under the microscope with a continuous CCD camera, the phase change process appears to happen in a fraction of a second, without visible nucleation process. The boiling begins from one or two channels first, and quickly develops into other channels. This behavior is different from that in macroscopic channels, where the phase change always begins from small bubble growth on the solid walls. This phenomenon will be further discussed in the following chapters.

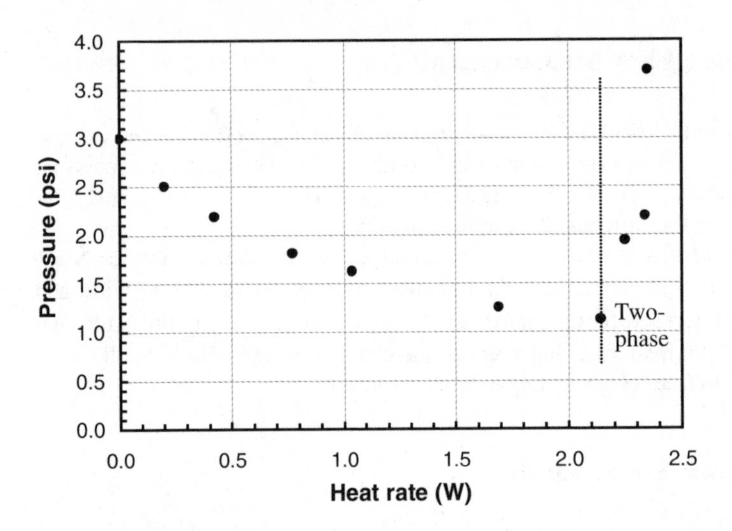

Fig. 4.5 Pressure change in Device No. 1, a 31 μm diameter, 40-channel device, measured as a function of heat rate for a constant DI water flow rate of 0.1 ml/min.

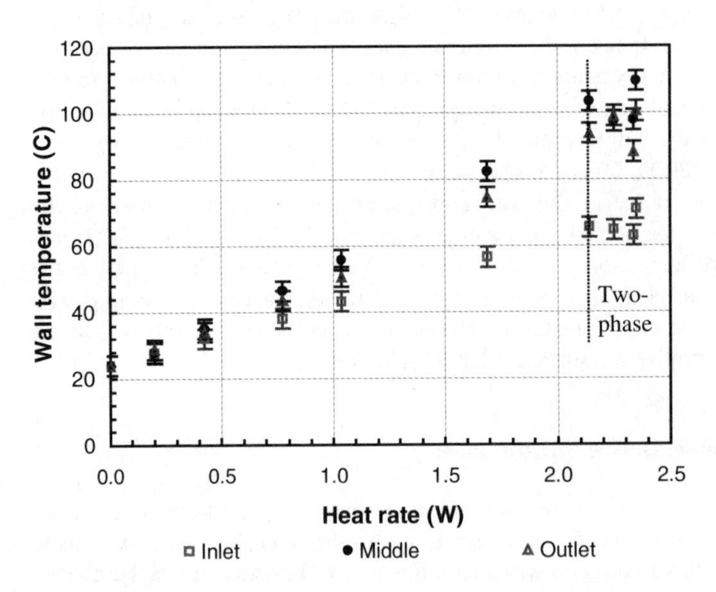

Fig. 4.6 Local wall temperatures of Device No. 1, a 31 μm diameter, 40-channel device, measured as a function of heat rate for a constant DI water flow rate of 0.1 ml/min.

Figures 4.5 through 4.7 show the pressure and local wall temperature changes as a function of increasing heat rate, as well as wall temperature distributions before and after the phase change. In Fig. 4.5, similar as in the heat sink chips, the pressure drop along the channels decreases with increasing power during the single-phase flow due to the decrease in water viscosity at higher temperatures. Immediately after the onset of boiling, because the density of vapor is much lower than that of water, the local volume flow rate in the channel suddenly increases. The resulting high velocity of the mixture yields a large pressure drop along the channels.

Local wall temperatures are plotted in Fig. 4.6. In the single-phase flow region, the wall temperatures increase linearly with the input heat rate, as expected from single-phase heat transfer theories. In the two-phase flow region, on the other hand, the data indicate a slight reduction in wall temperature at the initiation of boiling from all of the thermometers. An apparent wall temperature reduction is usually expected in pool boiling, but not in forced convective boiling. One interesting phenomenon is that, although the optical observation showed that boiling occurred at about the same place in each channel, the current and voltage signals at all thermometers began fluctuating in time immediately after the onset of boiling. The transient measurements in Chap. 5 further reveal that the

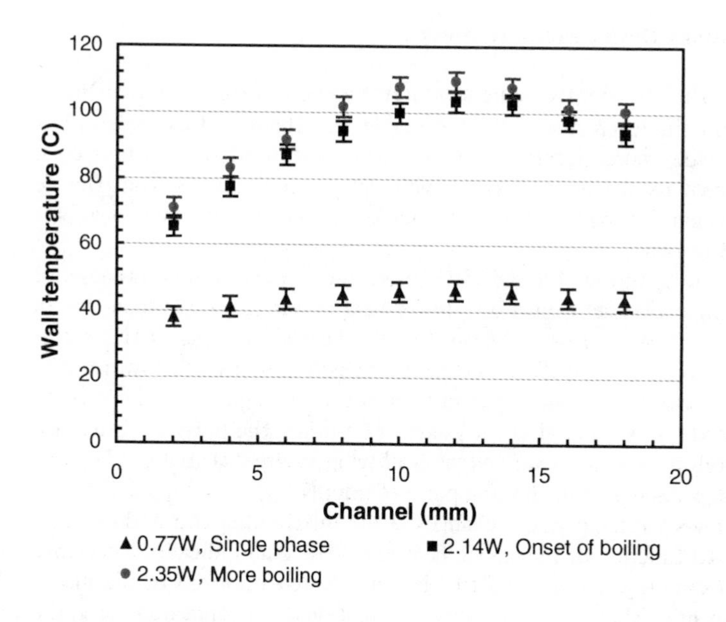

Fig. 4.7 Wall temperature distributions in Device No. 1, a 31 μm diameter, 40-channel device, measured at three heat rate levels for a constant DI water flow rate of 0.1 ml/min. The onset of boiling occurs at 2.14 W heat rate.

temperature reduction here is temporal. The data do not show temperature plateaus expected in the two-phase region, possibly due to partial dry-out in the channels after the fast nucleation process.

Figure 4.7 shows wall temperature distributions at various input heat rates. The measured local wall temperatures from the nine thermometers are plotted as a function of their locations along the channel. The profile in single-phase flow region is expected to be linearly increasing towards the exit of the channels. The measured parabolic shape indicates that there is conduction heat loss along the channel walls into the fluid entrance and exit regions, which is unavoidable in real measurement systems. This heat loss can be estimated from the thermal circuit model described in Chap. 3. Because the experiment deals with sub-cooled boiling, in which case the inlet water temperature is below the saturation temperature, the fully developed steady state boiling begins from the middle of the channels. Hence in the two boiling profiles, there is a linear region with wall temperatures below 100 °C, corresponding to the single-phase flow before the phase change occurs. In the two-phase region, however, the shape of the profile is less predictable because of the varying local flow conditions and associated heat transfer coefficients. Unlike the temperature-heat rate plots, here a temperature plateau does not necessarily appear due to the influences from local pressure and flow patterns.

4.3.3 Single-channel Device Measurements

Figures 4.8 through 4.10 plot the same measurements performed on Device No. 2, the 50 by 70 μm single channel. With a single test channel, Device No. 2 is expected to provide more accurate pressure and heat transfer measurements. These measurement results are compared with the previous results from Device No. 1, a multi-channel design, to find out whether the number of channels affects the experimental results.

The pressure curves in Fig. 4.8 and Fig. 4.5 are similar, except that the onset of boiling in the single channel occurs at a lower heat power (1.32 W) than that in multi-channels (2.14 W). This is because of the larger heat loss in the multi-channel design, which will be discussed in the modeling section. However, the same 0.1 ml/min water flow rate apparently causes a larger pressure drop in the single channel both before and after the onset of boiling. The pressure difference will affect the boiling process as well as the wall temperature distributions because of the pressure dependence of the boiling point of liquids.

Figure 4.9 shows the temperature changes at the inlet, outlet and middle point of the channel as a function of increasing heat rate. The data indicate no decrease in the wall temperature after the onset of boiling, which occurred in the multi-channel experiment. However, resistance fluctuations also appeared after the onset of boiling as in the multi-channel device and the earlier instrumented heat sinks. These transient fluctuations have 10-300 Hz frequencies and always accompany the phase change process. The local temperature profiles have a linear

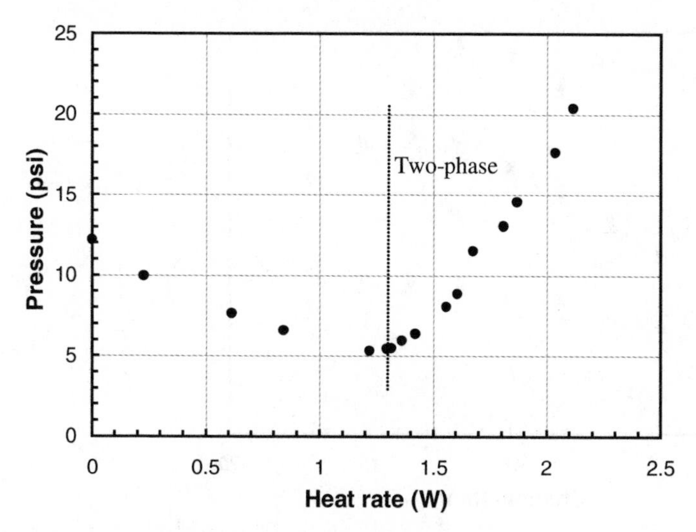

Fig. 4.8 Measurement of pressure change in Device No. 2, a 58 μm hydraulic diameter single-channel device, as a function of heat rate for a constant DI water flow rate of 0.1 ml/min.

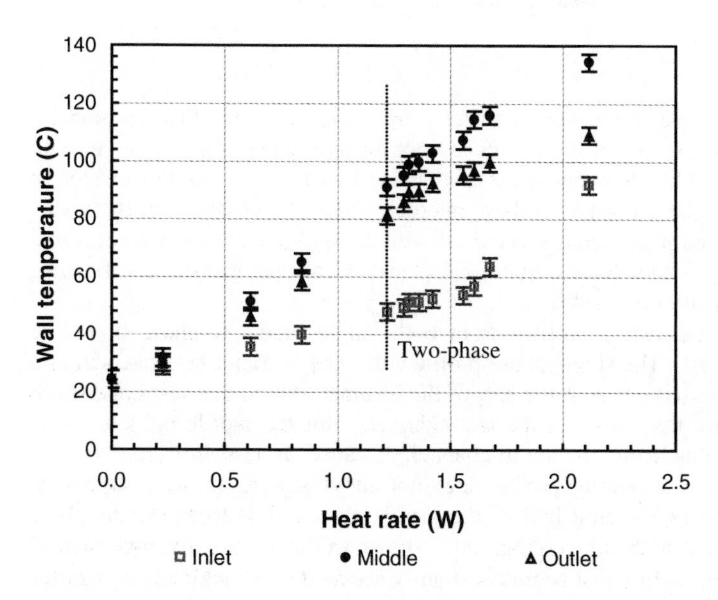

Fig. 4.9 Local wall temperatures of Device No. 2, a 58 μm hydraulic diameter single-channel device, measured as a function of heat rate for a constant DI water flow rate of 0.1 ml/min.

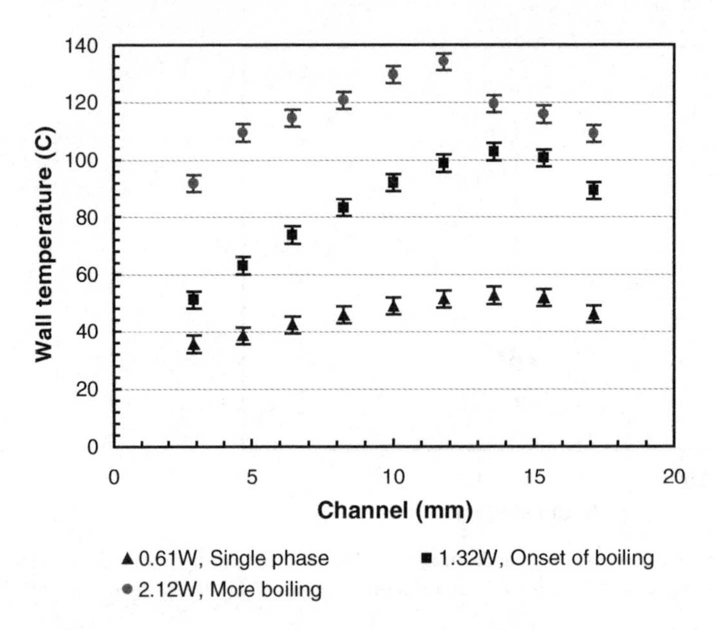

Fig. 4.10 Measurement of wall temperature distributions in Device No. 2, a 58 μm hydraulic diameter single-channel device, at three heat rate levels for a constant DI water flow rate of 0.1 ml/min. The onset of boiling occurs at 1.32 W heat rate.

region in the single-phase heat transfer. But the local thermal resistances, represented by the slopes of the curves, do not change much from single-phase to two-phase flow. This does not necessarily mean that the heat transfer cannot be enhanced by the phase change in these microchannels—as observed in the multi-channel device, the phase change occurred without visible nucleation process and this may lead to a flow pattern that yields lower local heat transfer coefficients, such as in partial dry-out conditions.

Three wall temperature profiles from both single- and two-phase flows are plotted in Fig. 4.10. The single-phase profile with 0.61 W input heat rate shows a slight temperature reduction at the end of the channel, due to the heat conduction loss to the fixture that supports the test channel. But the profile is more linear compared with that from the multi-channel, because of less heat loss in this device. The onset-of-boiling profile does not show significant wall superheat. The linear region in the first half of the profile corresponds to the single-phase flow in the channel with sub-cooling. Also shown in this profile, the highest wall temperature when boiling just begins is slightly above 100 °C, which is normal for nucleate boiling. At 1.32 W heat rate, when the boiling is much more intensive, partial dry-out conditions are responsible for the irregular shape in the third wall temperature profile.

4.3.4 Flow Patterns

From the temperature measurements, the phase change in these 28-60 μm diameter microchannels is not as stable as in macroscale channels. This is one impact of the small dimension of the microchannels to the bubble formation during the phase change, which will be discussed in Chap. 5.

As shown in Fig. 4.11, although the bubble nucleation process was not visible, optical microscopy indicates that the fully developed two-phase flow patterns in these 28-60 μm channels are mostly mixed liquid and vapor with a clear liquid-vapor interface fluctuating in position at high speeds. The measured wall temperatures suggest that typical nucleate boiling occurs in these plasma etched silicon microchannels.

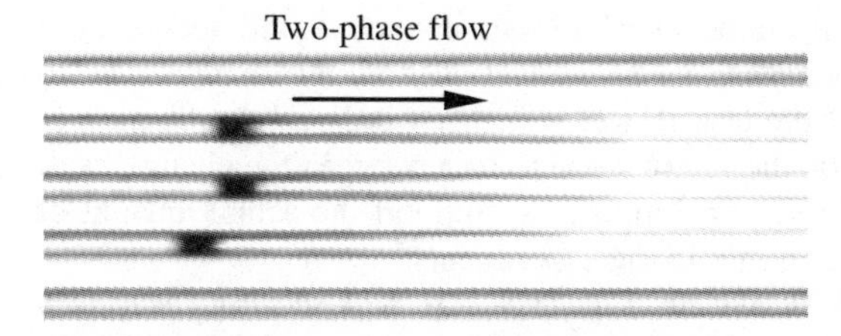

(a) The interfaces between the liquid and vapor phases in Device No. 1, with 20 μm wide, 70 μm deep (31 μm hydraulic diameter) channels.

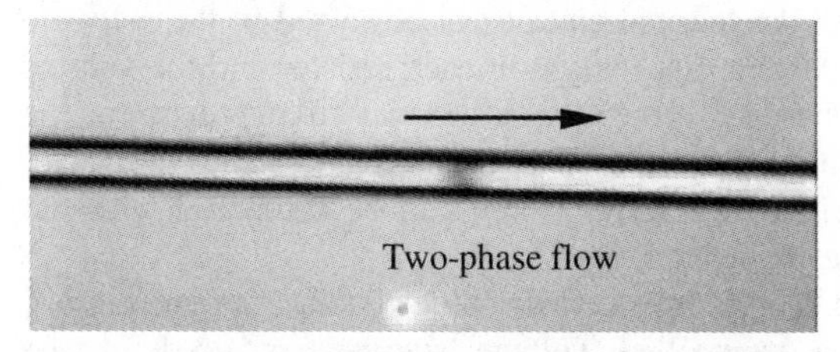

(b) The interface between the liquid and vapor phases vibrates at high frequencies in Device No. 2, a 50 μm wide, 70 μm deep (58 μm hydraulic diameter) channel.

Fig. 4.11 Images of two-phase flows in 31 μm and 58 μm diameter microchannels. The flow appears as a mixture of liquid and vapor with an interface between the single-phase and two-phase regions.

4.4 Modeling of Two-phase Internal Flow

Both two-phase flow and heat transfer models have been developed to simulate the experiments. The heat transfer model consists of macroscopic thermal resistances for the test system and microscopic finite volumes for internal microchannel flow. Heat distribution is taken care of in the thermal circuit model described in Chap. 3, which predicts the fraction of heat carried away by the microchannel flow. A detailed internal flow model that considers both single- and two-phase flows was developed to simulate the pressure and temperature distributions with finite volume method. By applying energy balance equations and using classical heat transfer correlations, the wall temperature profile has been simulated. The simulation and experimental results are in reasonable agreement.

4.4.1 Heat Loss Estimation

As indicated in the wall temperature profile measurements, the temperature reduction at the end of the channel indicates conduction from the channel wall to the entrance and exit regions of the test device. In other words, only a fraction of the applied heat rate is carried by the internal convection between the channel wall and the fluid. The purpose of this simulation is to find out the actual amount of the heat that was absorbed by the liquid in the experiments.

The thermal circuit model shown in Fig. 3.5 can be used for heat loss estimation. Table 4.2 lists the thermal resistance values of Device No. 2, a 50 μm wide and 70 μm deep single-channel design, at 0.1 ml/min water flow rate. Thermal conductivity of silicon is assumed 148 W/m-K [4.13].

With all the resistance values, the thermal circuit can be solved in the software package SPICE to simulate the current distribution in each resistor and the voltage at each node. Again, the currents and voltages in a thermal circuit represent the heat flows and temperatures in a thermal system respectively. The voltage at Node *61* simulates the exit fluid temperature, which can be compared with the measurement data to validate the simulation.

Since the two-phase flow is too complicated to model in the thermal circuit, single-phase flow (modeled as controlled voltage sources) is used in the simulation. The measured and estimated outlet water temperature during single-phase heat transfer experiments on Device No. 2 are plotted together in Fig. 4.12. With ±2.5 °C measurement uncertainty, the measurements and simulations are in good agreement, indicating that the model accounts well for the various heat transfer paths in the test system. It is true that the heat distribution will change in two-phase flow due to the convective resistance change as the result of a different convective coefficient; however this method provides the most reasonable estimation of the heat distribution. The model is then used to estimate the fraction of the heat carried away by the fluid during earlier experiments.

At 0.1 ml/min flow rate, the estimated system heat loss for Device No. 1 (the 40-channel design) is 39%, with 20% lost to pre-heating of inlet water; or, 61% of applied heat rate was absorbed by the internal microchannel flow. The system heat loss for Device No. 2 (the single-channel design) is 26%, with 12% lost to pre-heating; or 74% of the actual heat transfer happened in the microchannel.

Table 4.2 Thermal resistances of the single-channel device.

Resistance	Value (K/W)
R_w ($1/\dot{m}\, c_p$, fluid)	143.5
R1 ($1/hA$, reservoir)	13.8
R2 ($1/hA$, entrance)	44
R3 ($1/hA$, microchannel)	5.5
R4 ($1/hA$, exit)	44
R5 (removed in this design)	infinity
R21 (l/kA, entrance-reservoir)	80
R32 (l/kA, channel-entrance)	300
R34 (l/kA, channel-exit)	300
R45($1/kA$, exit-fixture)	55
R10 (fixture, estimated)	75
R50 (fixture, estimated)	75
R30 (free convection, estimated)	300

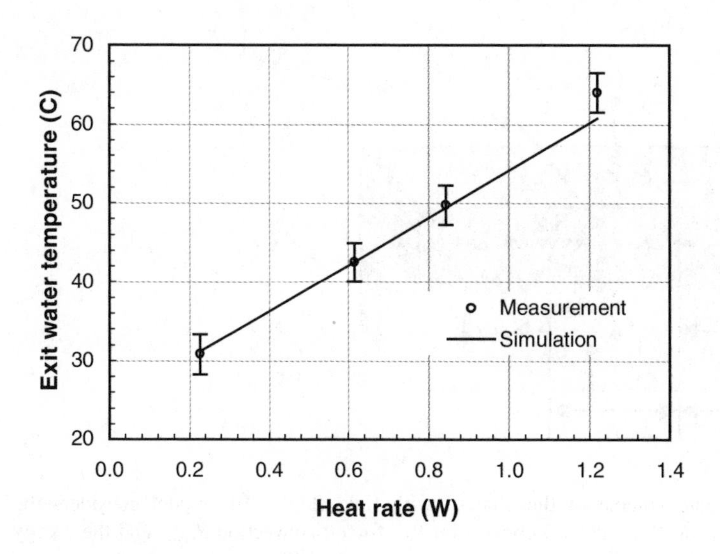

Fig. 4.12 Outlet water temperature measurement and simulation under single-phase heat transfer. The simulation is based on simple forced convective internal flow model.

As discussed in Chap. 3, the heat loss in the system is a function of the flow rate. As an example, at 0.5 ml/min flow rate, the system heat loss in Device No. 1 is as low as 5%, with 5% to pre-heating. However, increasing flow rate may be detrimental because the local temporal pressure increase during the phase change can be severe. Even silicon-glass bonds that survive 50 psi static pressure could be broken by the boiling, which has been observed in experiments. Hence for safety reasons, the flow rate in the experiments was limited to 0.1 ml/min. The cause of the "explosive boiling" will be discussed in detail in Chap. 6.

As an additional comment, given the substrate thickness 400 μm and the channel depth 50-100 μm, the maximum distance between the thermometers and the bottom of the channel is 350 μm. The thermal resistance for conduction through the silicon in this distance is approximately 0.23 °C/W (single-channel) or 0.06 °C/W (multi-channel), while the convection resistance into the fluid is never smaller than 3.8 °C/W. Therefore, the thermometers provide an accurate measurement of the wall temperature.

4.4.2 One-dimensional Finite Volume Model for Two-phase Internal Flow

A one-dimensional microchannel flow and heat transfer simulation has been developed for comparison with the experimental results [4.14]. The simulation numerically solves energy equations for heat conduction in the silicon wall and

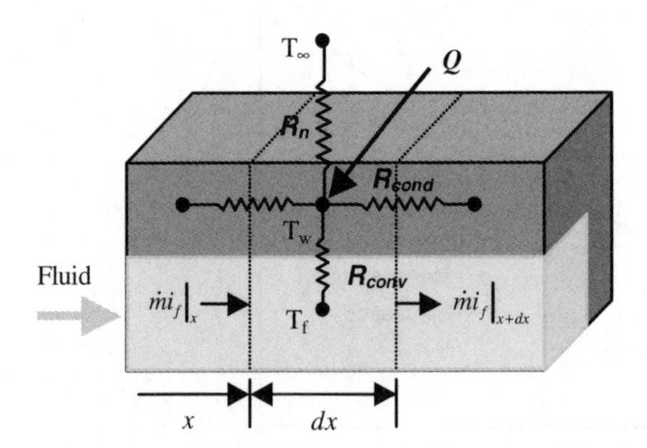

Fig. 4.13 A finite volume of the channel with fluid flow. The model considers the conduction in silicon R_{cond}, natural convection R_{nc}, forced convection R_{conv}, and the energy carried by the fluid $\dot{m}\, i_f$. The environment temperature T_∞, silicon wall temperature T_w, and the fluid temperature T_f are treated as nodes between the thermal resistors. By courtesy of Jae-Mo Koo.

convection by the fluid, with boundary conditions dictated by the heat loss to the environment. The simulation uses the finite volume method and considers the temperature and pressure dependence of the liquid and vapor properties based on correlations to tabulated data.

As shown in Fig. 4.13, the simulation is one-dimensional in the flow direction along the channel and the average local wall temperature T_w and the fluid temperature T_f, are used in each control volume. The governing energy equations are:

$$\frac{d}{dx}\left(k_w A_w \frac{dT_w}{dx}\right) - \eta h_{conv} p(T_w - T_f) - \frac{w(T_w - T_\infty)}{R_{env}} + q''w = 0 \tag{4.1}$$

$$\dot{m}\frac{di_f}{dx} - \eta h_{conv}\, p(T_w - T_f) = 0 \tag{4.2}$$

where x is the coordinate along the channel, A_w is the channel wall cross-sectional area, p is the perimeter of the channel cross section, and w is the pitch of one channel. The fin effectiveness, η, accounts for the temperature variation within the side walls. k_w is the thermal conductivity of silicon, $\dot{m}$ is the mass flow rate, and h_{conv} is the convection coefficient for heat transfer between the channel walls and the fluid. Kandlikar's correlation was used in the modeling because it had been proven to fit a broad range of data for flow boiling:

$$h_{conv} = h_l[C_1 C_O^{C_2}(25Fr_{le})^{C_5} + C_3 Bo^{C_4} F_K] \tag{4.3}$$

where the two-phase convection coefficient h_{conv} is predicted by multiplying a factor to the single-phase liquid convection coefficient h_l. The dimensionless parameters C_0-C_5, Froude number Fr_{le}, Bo and F_K are factors determined by the flow conditions [4.4]. The fluid enthalpy per unit mass, i_f, for two-phase flow is expressed in terms of local fluid quality γ, which is the mass fraction of the vapor phase, using

$$i_f = (1-\gamma)i_l + \gamma i_v \tag{4.4}$$

where subscripts l and v refer to liquid and vapor phase in two-phase flow, respectively. Equation (4.1) accounts for heat conduction along the channel wall in the first term, convection heat transfer in the second term, and the natural convection heat loss to the environment using the resistance R_{env} in the third term. Radiation is neglected due to its very small magnitude compared with other terms. The fluid flow equation (4.2) relates the change of the average enthalpy density of the fluid to the heat transfer rate into the fluid from the channel walls.

The pressure distribution is calculated from the flow model—the homogeneous model or the annular flow model. With the homogeneous model, the pressure distribution is governed by

$$-\left(\frac{dP}{dx}\right) = \frac{fm''^2}{2\rho D_H} + \frac{d}{dx}\left(\frac{m''^2}{\rho}\right) \tag{4.5}$$

and with the annular model, the pressure is determined by

$$-\left(\frac{dP}{dx}\right)=\frac{2\tau_i}{D_H/2-\delta}+\frac{m''^2}{\alpha}\frac{d}{dx}\left(\frac{x^2}{\alpha\rho_v}\right) \tag{4.6}$$

where ρ is the density of the liquid-vapor mixture, ρ_v is the density of vapor phase, f is the globally averaged friction factor, and D_H is the channel hydraulic diameter. The mass flux m'' is related to the mass flow rate $\dot{m}$ by $m''=\dot{m}/A_c$, where A_c is the cross sectional area of the flow channel. δ is the liquid film thickness, τ_i is the shear stress at the liquid-vapor interface in the annular flow model, and α is void fraction, which is the ratio of the vapor flow cross sectional area to the total flow cross sectional area. The friction factor f and the convection coefficient are taken from experimental correlations and are discussed in detail in [4.14].

The data of Stanley et al. for heat flux and friction coefficient for two-phase flow along channels of comparable dimensions lend more support to the homogeneous flow model [4.6]. From the flow pattern visualizations in Fig. 4.11, the homogeneous model is also supported. However, for comparison purpose, both models are simulated and the results are discussed in the following section.

4.5 Discussion

The experimental data from the single-channel device are compared with both simulation results from homogenous and annular flow models. The simulation parameters are given in Table 4.3.

Figure 4.14 shows reasonably good agreement between simulations and measurement data for the pressure drop as a function of applied heat rate, particularly for the homogeneous model. The decrease of the pressure drop with increasing heat rate below 1.32 W results from the decreasing viscosity of the pure liquid phase. The calculations under-predict the pressure drop in this regime due to fluid pre-heating in the inlet reservoir. The pressure increase with increasing power in the two-phase region results from the high mixture velocity caused by

Table 4.3 Two-phase flow and heat transfer simulation parameters.

Parameter	Value
k_w (wall thermal conductivity)	148 W/m-K
A_w (wall cross sectional area)	3500 μm^2
P (perimeter of channel cross section	240 μm
W (freestanding beam width)	500 μm
Q_v (volume flow rate)	0.1 ml/min
D_H (channel hydraulic diameter)	58 μm

liquid rapidly evaporating into the vapor phase. The annular flow model yields less pressure drop than the homogeneous model because it does not consider the mixing effect on the pressure drop.

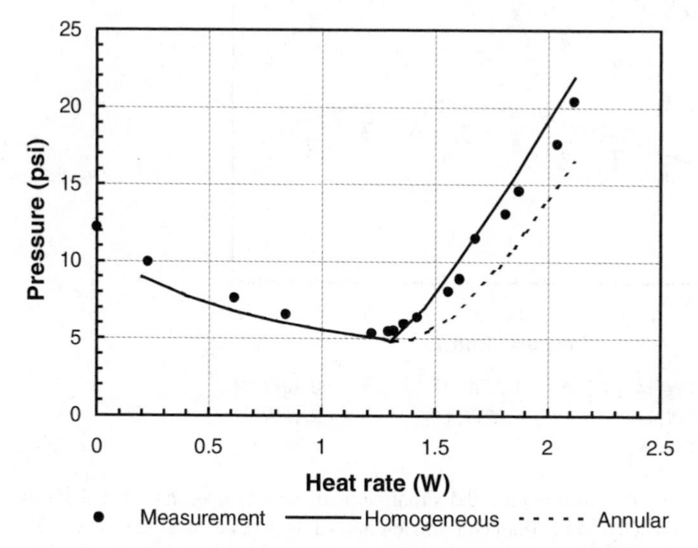

Fig. 4.14 Comparison of measurements and simulations of pressure change in Device No. 2, a 58 μm hydraulic diameter single-channel device, as a function of heat rate for a constant DI water flow rate of 0.1 ml/min. The homogeneous model agrees better with the experimental results than the annular flow model. In collaboration with Jae-Mo Koo and Linan Jiang.

Figure 4.15 compares simulations and experimental data for the wall temperature distributions along the channel. Relatively poor agreement exists in the middle of the channel for 0.61 W heat rate and the entry region for 2.12 W heat rate. The deviation for 0.61 W results from an estimated value of the environmental thermal resistance, R_{env} in Equation (4.1). The deviation for 2.12 W is very likely due to the effects of liquid pre-heating. The simulations are reasonably effective at predicting the temperature magnitude and the onset of boiling. Since the pressure field evaluated by the annular flow model is lower than that of the homogeneous model, its corresponding saturated-temperature field is also lower than that of the homogeneous model. Although the temperature simulations do not provide sufficient detail to choose between the annular and homogeneous flow models, the pressure drop simulations lend support to the homogeneous model.

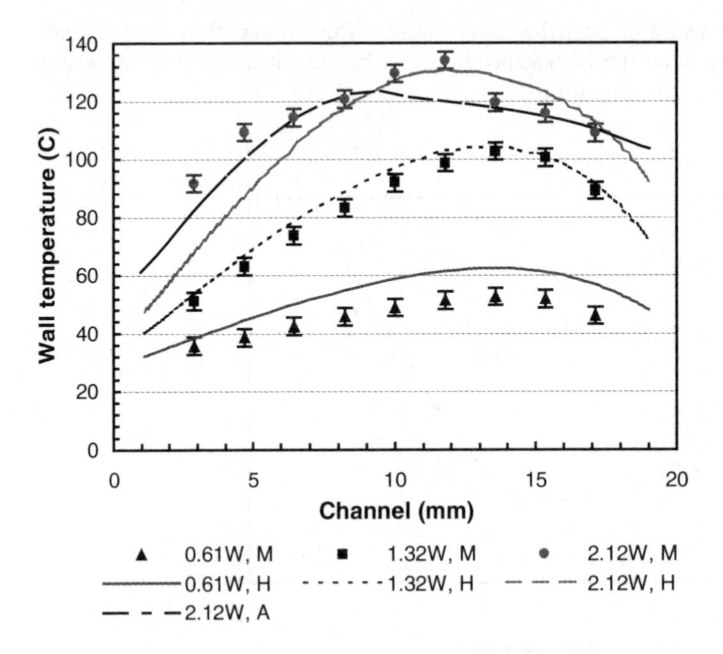

Fig. 4.15 Comparison of measurements and simulations of wall temperature distributions in Device No. 2, a 58 μm hydraulic diameter single-channel device, at three heat rate levels for a constant DI water flow rate of 0.1 ml/min. The onset of boiling occurs at 1.32 W input heat rate. In the legend, M refers to measurements, H refers to simulation results from the homogeneous flow model, and A refers to simulation results from the annular flow model. In collaboration with Jae-Mo Koo and Linan Jiang.

Both the modeling results and the preliminary visualizations of the phase change in 28-60 μm microchannels suggest that homogeneous liquid-vapor mist flow is the dominant two-phase flow regime. The liquid and the mist are separated by an interface oscillating in position at high speeds, and after a very short annular flow region, the vapor and liquid mixture moves at the approximately the same velocity. The qualitative agreement between the experimental and simulation results support the homogeneous flow model as well as Kandlikar's convection correlation in the heat transfer modeling. And this is the first time that Kandlikar's correlation is proven to function in channels with hydraulic diameters below 500 μm.

5 Boiling Regimes and Transient Signals Associated with the Phase Change

5.1 Background: Boiling Regimes in Large and Small Channels

Since the heat transfer coefficient associated with the phase change in microchannels is significantly affected by the two-phase flow regime, visualization of the phase change in microchannels has always received research interests. In this chapter, questions such as how the boiling regimes in microchannels differ from those observed in large tubes, and what exactly happens at the moment of the phase change will be answered.

5.1.1 Nucleate Boiling in Large, Horizontal Tubes

In a large, horizontal heated tube, the two-phase flow after the onset of boiling can be roughly divided into four regimes, namely, bubbly flow, plug/slug flow, annular flow and mist flow [5.1]. The bubbly flow appears at the beginning of the phase change, with small bubbles growing and detaching from the tube walls. Later small bubbles coalesce to form plug and slug of vapor, called plug/slug flow. At higher heat fluxes and flow velocities, annular flow appears in the form of a liquid film moving along the inner surface of the tube and the vapor moving through the core at a larger velocity. When the vapor quality keeps increasing, the liquid finally exists as droplets appearing in the vapor core, referred to as the mist flow. Due to the buoyancy force, the vapor in a macroscale tube tends to migrate to the top of the tube and the bottom carries more liquid. The flow regimes and corresponding convective heat transfer coefficients are shown in Fig. 5.1. In the fully developed single-phase laminar flow, the convective coefficient is a constant independent of the flow rate. The number increases sharply at the bubbly, plug and slug flow regions, reaches its maximum at the annular flow regime, and decreases after the mist flow forms.

5.1.2 Boiling in Microchannels—Previous Research

Boiling regimes in microscale channels have been reported to be different from those in large tubes. In summary, the bubbly and plug/slug flow regimes seem to be absent in mini- and microchannels below 1 mm diameter.

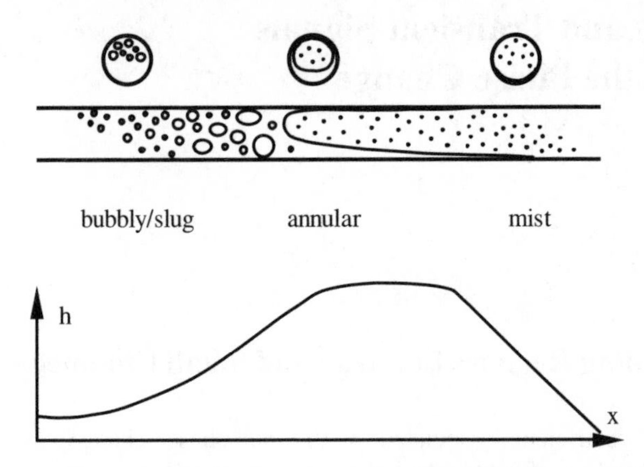

Fig. 5.1 Two-phase flow regimes and heat transfer coefficients in a macroscale horizontal tube. The phase change initiates from a bubbly flow regime; turns into plug and slug flow with larger vapor bubbles; then develops into annular flow where a high velocity vapor core is wrapped with a thin layer of liquid; and the flow eventually becomes a mist where the vapor and liquid droplets move at the same velocity [5.1].

Jiang et al. studied water-vapor phase change in 26 μm and 53 μm V-shaped microchannels with optical access [5.2]. Bubble nucleation in large manifolds to the flow channels and annular flow regime were observed at low to medium heat fluxes. However, the increase in the heat transfer coefficient during the phase change was not detected.

Peng et al. conducted phase change experiments in 0.6×0.7 mm stainless steel channels [5.3]. They reported that the heat transfer in the single-phase regime was intensified and no partial boiling or nucleation was observed. The fully developed boiling seemed to appear suddenly at certain unusually high heat fluxes.

Kenning and Yan designed a single vertical rectangular channel with 2×1 mm cross section and 248 mm length, which allowed simultaneous wall temperature measurement, flow regime visualization, and transient pressure measurement [5.4]. They recorded bubble growth in the channel and successfully detected pressure pulses caused by the acceleration of liquid slugs resulting from confined bubble growth.

Peles et al. studied the phase change in 50-200 μm silicon channels [5.5]. They proposed that explosive boiling (a sudden phase change without visible bubble growth) might occur at small Pelcet numbers, which lead to fine bubble growth, blockage of the channel, and explosive boiling.

Bubble nucleation has been visualized in both larger than 1 mm diameter channels and large areas such as manifolds in microchannel test devices. However, the much smaller channels, where an "explosive" style boiling is more often reported, are of more interest to the study of microchannel heat sinks. This chapter presents phase change visualizations that lead to the answers of how it occurs and how the two-phase flow develops in microchannels below 150 μm diameter. Some characteristic transient phenomena associated with the phase change in this dimension are also discussed in this chapter.

5.2 Design Parameters of Test Devices

The design of the new microchannel devices considers implementation of automatic data acquisition for transient state measurements. Aluminum heaters and doped single crystal silicon thermometers are integrated on the test chip, as described in Sect. 3.3.2.

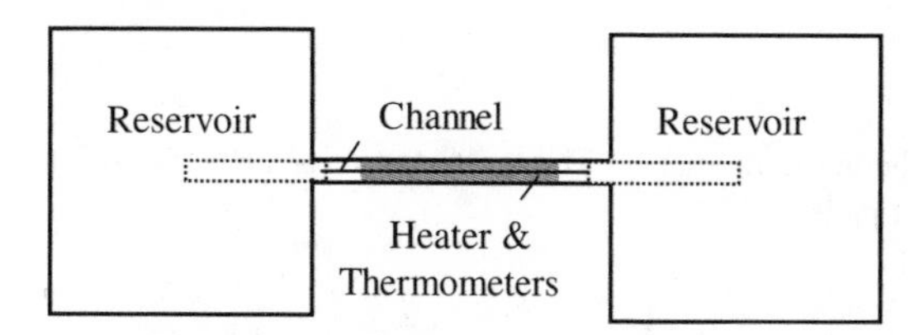

Fig. 5.2 Schematic of the test device. The aluminum heater and silicon thermometers are integrated on the back side of the freestanding silicon beam, with a single test channel formed on the front side. The channel is sealed with a glass cover slide.

Figure 5.2 shows the plan view of the test device. The overall dimension is 2×6.5×0.05 cm. The freestanding beam is 1.3 mm in width. Two reservoirs are formed on the two ends of the microchannel, and each chip has a single test channel. The channel hydraulic diameters range from 27 to 171 μm. The structural parameters are listed in Table 5.1.

Table 5.1 Structural parameters of single test channels with separate heaters and thermometers.

Structure	Dimension
Reservoirs	600 μm wide, 250 μm deep, 1 cm long
Freestanding beam	1.3 mm wide, 2 cm long
Channel dimensions	20-150 μm wide, 40-200 μm deep, 2 cm long
Aluminum heater	50 μm wide, 2 cm long, 2 μm thick, 10 lines resistance 50 Ω
Resistive thermometers	150 μm long, 10 μm wide, 1 μm junction depth resistance 8 kΩ
Si substrate thickness	480 μm
Glass thickness	520 μm

Seven thermometers are formed along the 2 cm long channel. The schematic of the thermometers is shown in Fig. 5.3. The overall size of the thermometer is 110×150 μm with two 200×200 μm contact pads. Because of their small size, the thermometers provide very localized wall temperature measurements.

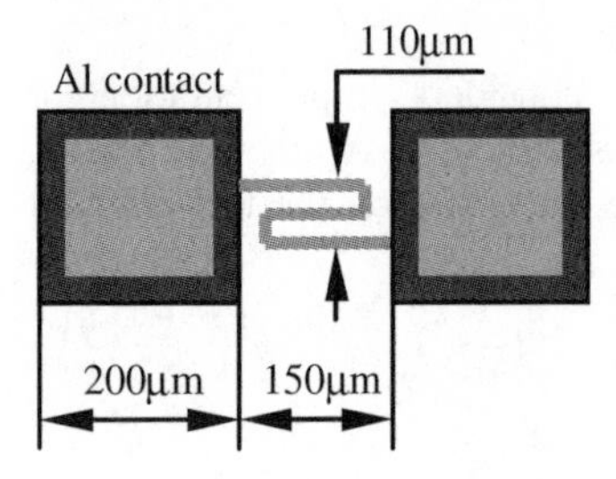

Fig. 5.3 Schematic of a doped silicon thermometer. The thermometer is 110×150 μm in size, with two 200×200 μm contact pads.

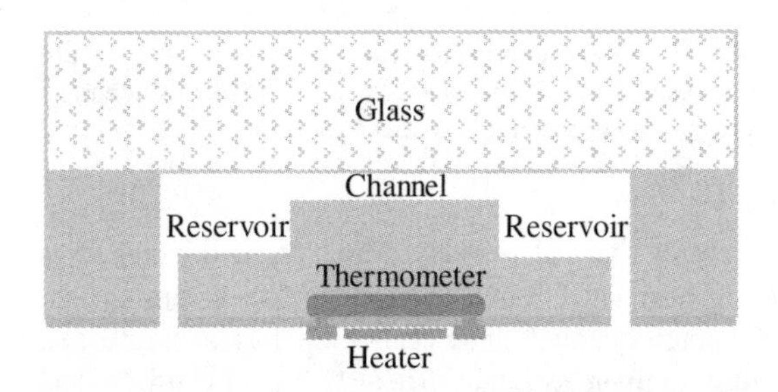

Fig. 5.4 Cross sectional view of the test channel with one aluminum heater and seven doped silicon thermometers.

Figure 5.4 shows the cross sectional view of the test device. The fabrication process flow chart is listed in Appendix B. Two images of the front and back sides of the microchannel with the heater and thermometers are given in Fig. 5.5. As suggested in Chap. 3, a major design consideration for the aluminum heater is the current density, which must be less than the limit of electromigration that can melt the wire. In this design, the aluminum heater is made of 10 aluminum lines, each of which is 50 μm wide and 2 μm thick, with a total of 50 Ω resistance. At 10 W heat rate, the current is 0.46 A; hence the current density is 0.46 A/ (50 μm × 2 μm) = 4.6×10^9 A/m^2, about a half of the limit density of 10^{10} A/m^2.

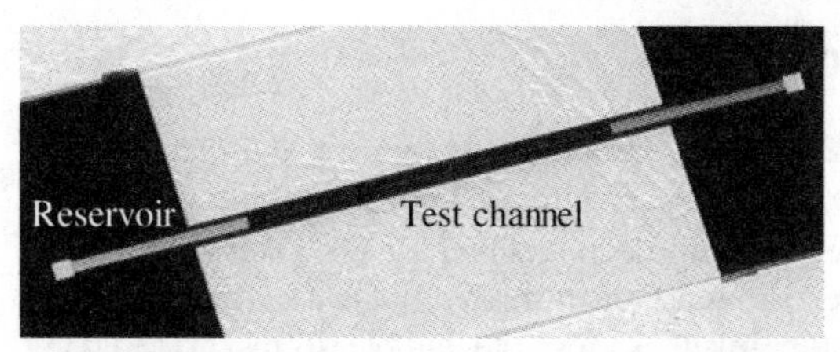

(a) Front side of the device with a single test channel.

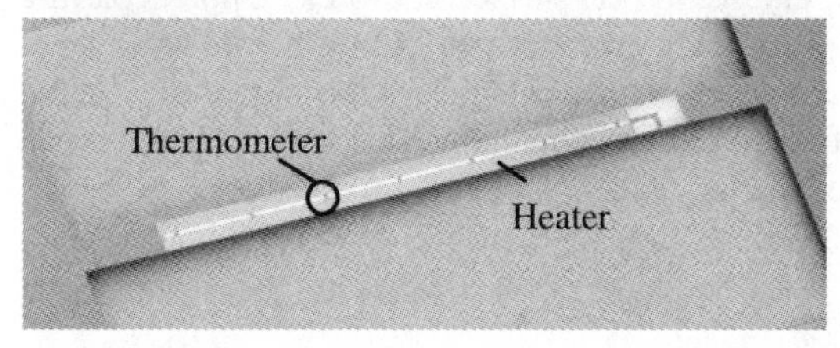

(b) The aluminum heater and silicon thermometers on the back side.

Fig. 5.5 Images of the instrumented silicon single-channel devices.

5.3 Visualization of Nucleate Boiling in Microchannels

All previously discussed two-phase experiments were conducted in microchannels with less than 60 μm hydraulic diameter. Although the steady state wall temperature measurements suggest that the boiling mechanism falls in nucleate boiling from the wall superheat, it has never been optically confirmed that bubbles form and grow in the microchannels. Only the interface between the liquid and the vapor phase fluctuating at high speeds has been observed in the fully developed boiling flow.

In the earlier experiments, a continuous CCD camera without a shutter was used for image captions. From the discussions in Sect. 3.4.1, with objective lens NA values of 0.1 (4X), 0.25 (10X), and 0.4 (20X), and 30 frames/s camera frame rate (integration time t=0.033s), the maximum allowed object speed is 187 μm/s at 4X magnification. This explains why the interface images in Fig. 4.11 are not clear. A SONY 2735 CCD camera was selected for image captions in the new experiments. The camera has a built-in shutter that automatically controls the light integration time from 1/50s to 1/100000s according to the light intensity. With t=1/50 to 1/1000s actual integration time, the camera is capable of clearly capturing objects that move at 78-1555 μm/s speed with 20X magnification. The frame rate of the camera is 30 frames per second.

5.3.1 Bubble Nucleation in Microchannels

With the shutter controlled CCD camera, bubble nucleation on the channel wall has been clearly recorded for the first time. Figure 5.6 shows a bubble growing on the side wall of a 250 μm wide, 130 μm deep channel, at DI water flow rate of 0.5 ml/min. The series of frames show a typical bubble nucleation and departure process in microchannels. In the image series, the bubble forms at *t*=0.033s, grows to around 80 μm diameter, and leaves the surface at *t*=0.2s. Upon departure of the bubble from the side wall, it quickly grows larger and turns into a transient annular flow, leaving a thin liquid layer on the channel walls. After the bubble exits the channel, the liquid refills the entire channel. This cycle repeats when each bubble grows and leaves the channel wall.

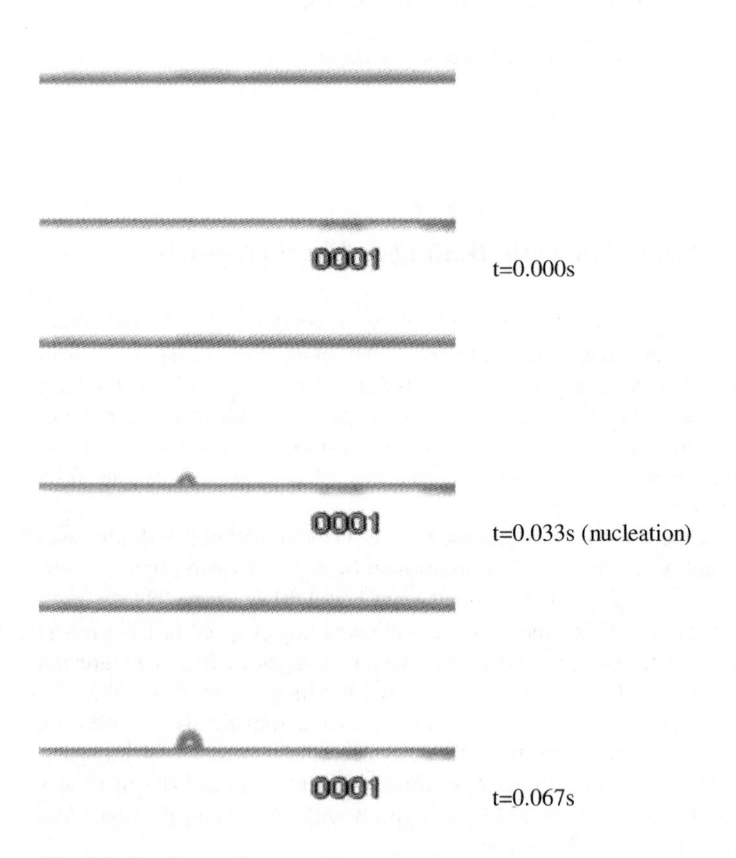

Fig. 5.6 Caption on the opposite page.

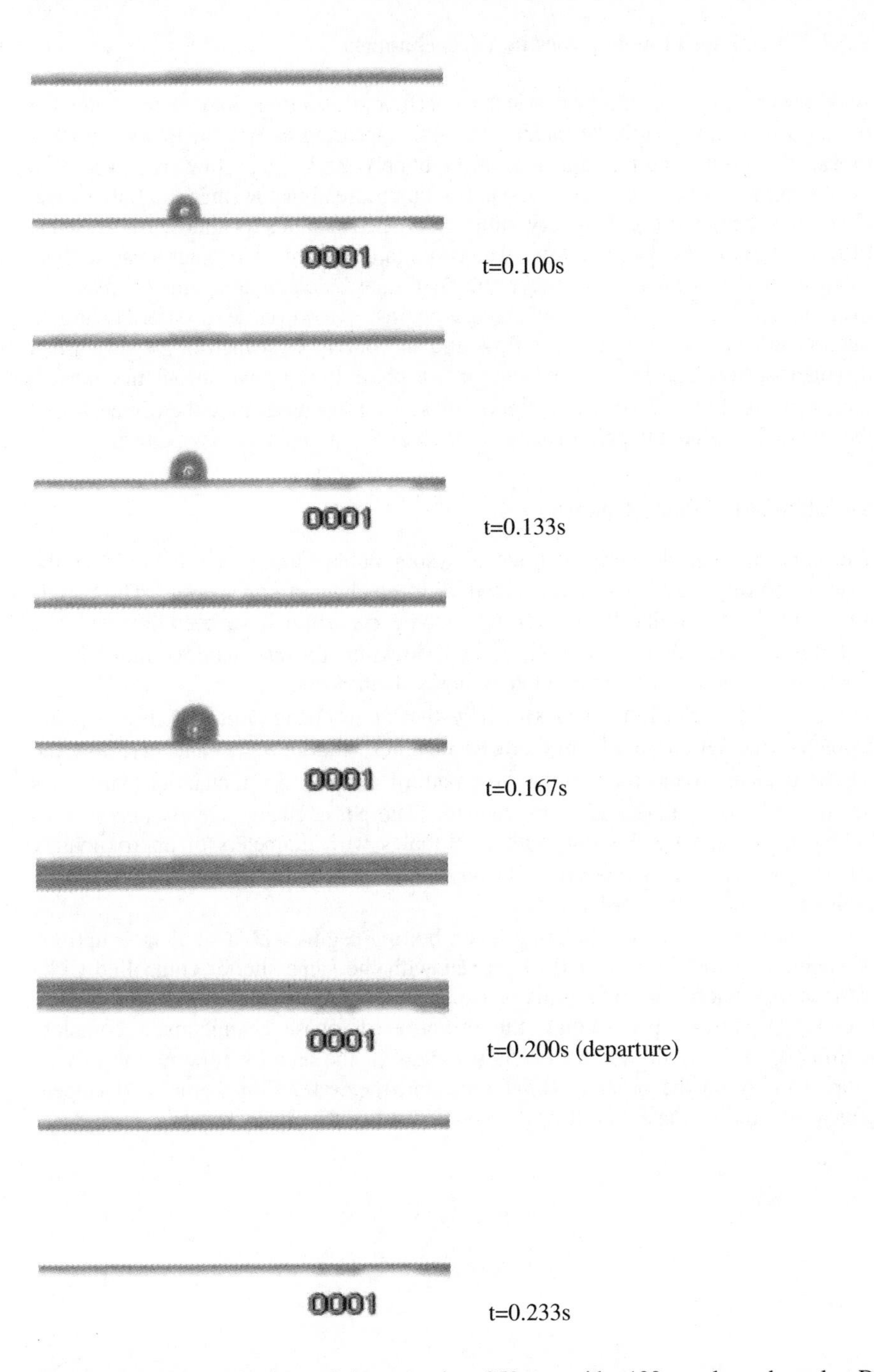

Fig. 5.6 Bubble nucleation and departure in a 250 μm wide, 130 μm deep channel at DI water flow rate of 0.5 ml/min.

5.3.2 Two-phase Flow Regimes in Microchannels

As shown in Fig. 5.1, the heat transfer coefficient of a two-phase flow is dictated by the flow regime, with the maximum value appearing in annular flows. In large tubes, this flow regime appears after bubbly and slug flow regimes. In microchannels, however, many research results already show that the bubbly and slug flow regions are either very short or absent, leading to almost immediately formed annular flow or mist flow. Hence a major advantage of microchannel heat sinks is that it is possible to make the heat transfer coefficient quickly reach its maximum value after the phase change begins. However, excessive boiling in microchannels also leads to mist flow and dry-out spots, which negatively affect the internal heat transfer. By presenting two-phase flow visualizations in channels ranging from 27 to 171 μm diameters, this section discusses how the dimension of the channel impacts the phase change as well as the two-phase flow pattern.

Initiation of the Phase Change

The nucleation mechanism, or how the vapor bubble forms in the liquid at the onset of boiling, seems to be associated with the channel dimensions. The bubble nucleation and annular flow after the bubble departure have been imaged in a 171 μm diameter channel in Sect. 5.3.1. However, for less than 50 μm diameter channels, the boiling occurs in a much more violent form.

Figure 5.7 shows four images of how the phase change initiates in a 113 μm diameter channel and a 44 μm diameter channel. Figure 5.7(a) and (b) show the bubble nucleation and the annular flow pattern in the 113 μm channel, which has much similarity as in the 171 μm channel. The phase change clearly begins with bubble nucleation on the side walls. Bubbles with diameters of approximately 10-100 μm have been imaged. The period of bubble nucleation and departure cycles is on the order of 0.1-0.2 s.

In a 44 μm channel, on the other hand, boiling begins with a sudden "eruption" of vapor, followed by a mist flow. Even with the same shutter controlled CCD camera, the bubble growth process could not be recorded. Since the camera captures 30 frames per second, the initiation of phase change must complete within 0.033s. As shown in Fig. 5.7(c) and (d), the annular flow region is very short, usually on the order of 0.5-1 mm, and the water film seems to disappear afterwards, indicating a mist flow or even dry-out in the channel.

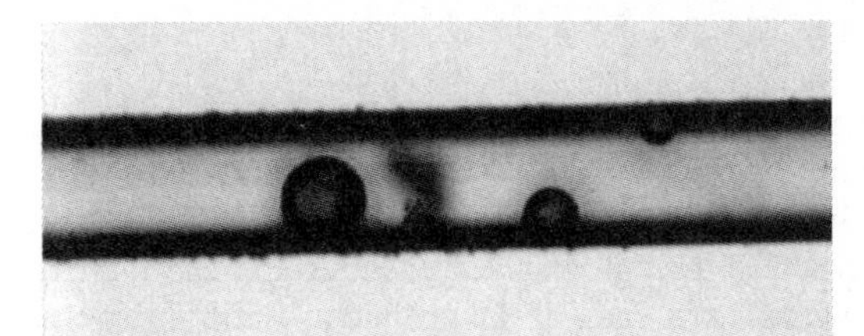

(a) Bubble nucleation in a 100 μm wide, 130 μm deep, 113 μm diameter channel.

(b) Annular flow in the 113 μm diameter channel.

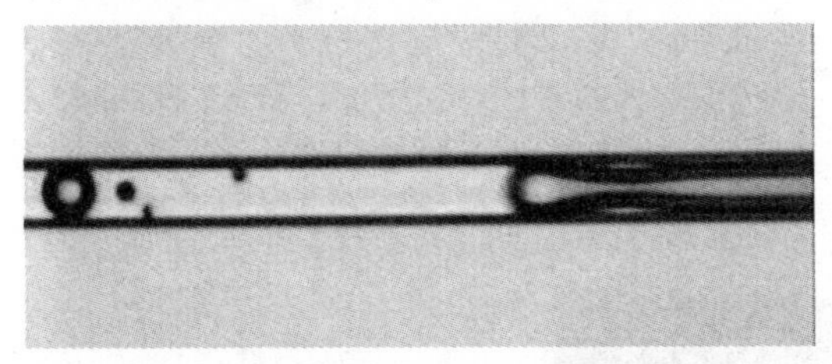

(c) Bubbles immediately turn to annular flow in a 50 μm wide, 40 μm deep, 44 μm diameter channel.

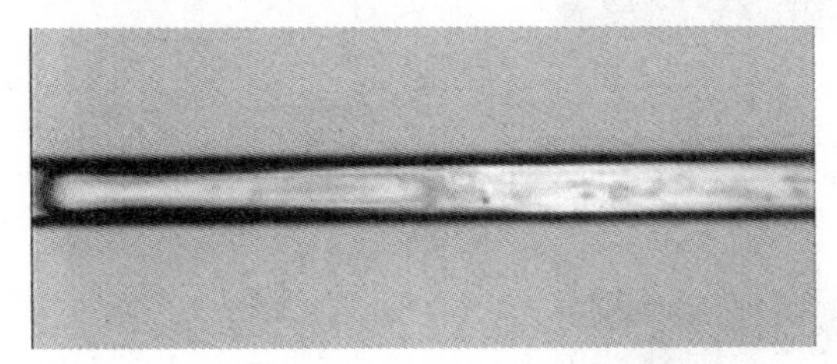

(d) Mist flow forms after a very short annular flow region in the 44 μm diameter channel.

Fig. 5.7 Phase change mechanisms in microchannels. Boiling initiates from bubble growth and the annular flow is the dominant two-phase flow regime in a 113 μm diameter channel. While in a 44 μm diameter channel, boiling initiates in an eruption style, and the mist flow is the dominant flow regime.

Flow Pattern Development

The two-phase flow patterns in microchannels are closely related to the nucleation mechanism, therefore also affected by the channel dimensions. Figure 5.8 shows two images of the liquid-vapor interfaces in the 113 μm channel. Because of the small confined space in microchannels, a single bubble consumes the entire cross section and is forced to grow towards both ends of the channel. The front of the bubble moves against the upstream liquid due to the sudden increase of the volume. Once the downstream end of the bubble reaches the exit, the vapor discharges to the reservoir, and the liquid pushes the rest of the vapor bubble out and the channel is refilled with liquid. At low heat fluxes, the single-phase flow and the transient annular two-phase flow transits between each other whenever a bubble forms and departs the channel wall, similar as the sequence of images shown in Fig. 5.6. At higher heat fluxes, the bubble growth-departure period becomes shorter and more bubbles leave the channel wall simultaneously; as a result, a steady annular flow appears with varying liquid layer thickness along the channel.

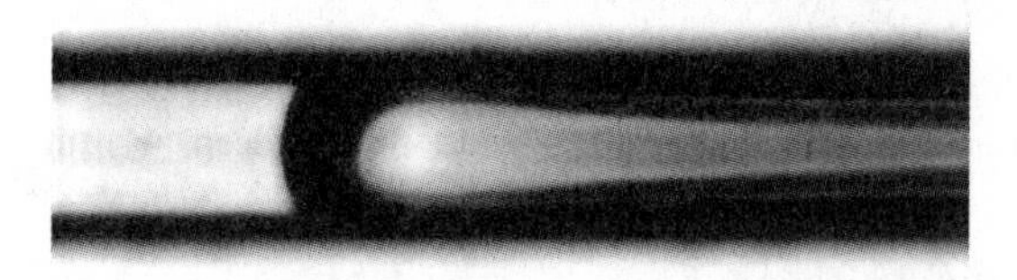

(a) The liquid-vapor interface at the inlet of the channel.

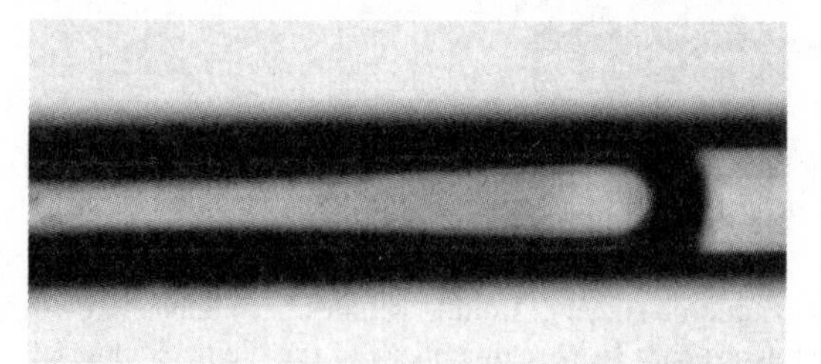

(b) The liquid-vapor interface at the outlet of the channel.

Fig. 5.8 Close-up images of the liquid-vapor interface in a 100 μm wide, 130 μm deep, 113 μm hydraulic diameter channel.

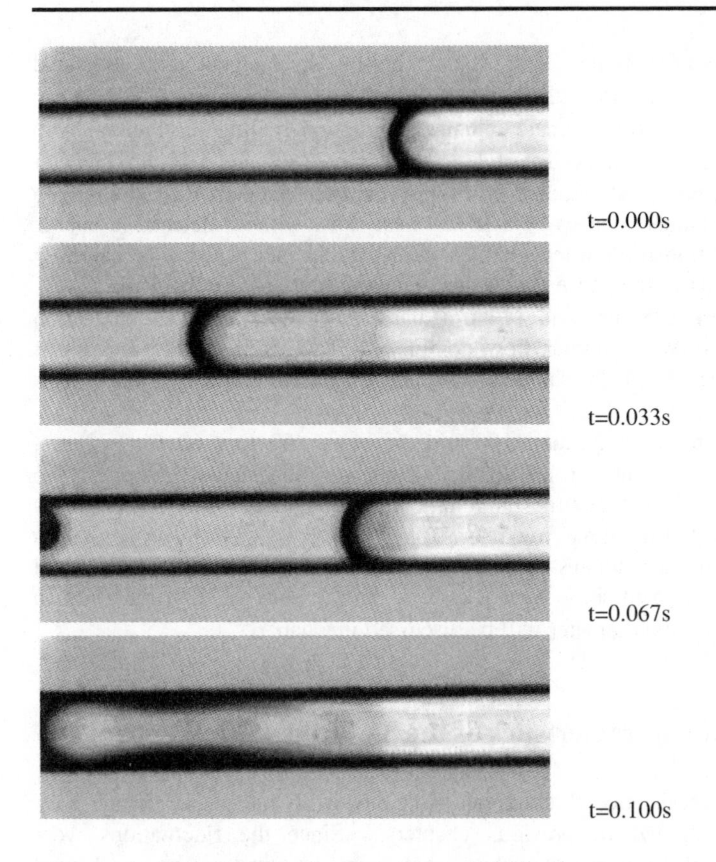

Fig. 5.9 The liquid-vapor interface oscillation in a 50 μm wide, 40 μm deep, 44 μm hydraulic diameter channel.

In Fig. 4.11(b), a dark shade separating the liquid and vapor phase in a 58 μm diameter channel is shown, which is an interface oscillating in position at high speeds. The same phenomenon also appears in the new 44 μm diameter channel. With the shutter controlled CCD camera, the motion of the interface has been clearly recorded and is shown in Fig. 5.9. Unlike in the 113 μm diameter channel, bubble growth and annular flow patterns cannot be captured. Instead, an oscillating liquid-vapor interface forms directly as the result of the eruption boiling and fast vaporization.

Discussion

Extensive experiments have been carried out in all channels with hydraulic diameters ranging from 27 to 171 μm at various flow rates. Optical observations confirm that the bubbly and plug/slug flow regimes are absent in all channels due to the small internal volume. Two distinct boiling mechanisms and associated flow regimes exist in sub-150 μm diameter channels:

In smaller than 50 μm diameter channels, eruption boiling mechanism with mist flows dominates. This boiling style appears in a 27 μm and a 44 μm diameter channels and the previously studied 28-60 μm diameter channels. The two-phase flow in these channels behaves as a steady mist flow, without detectable bubble growth or flow transition after bubbles exit the channel. Even at carefully controlled heat fluxes, the phase change occurs under the same mechanism.

In the 171 μm, 138 μm, and 113 μm diameter channels, normal nucleation boiling mechanism with annular flow dominates. The phase change appears as bubbles first depart from the channel wall, and then form a steady annular flow inside the channel.

The channels between 50 and 100 μm diameters (52 μm, 60 μm, 72 μm, 88 μm, and 97 μm) show a transitional, but not very repeatable behavior. The two-phase internal flow sometimes is annular flow dominant, while sometimes is mist flow dominant at the same flow rate and heat flux.

The nucleation mechanisms and flow patterns significantly affect the wall temperature and the two-phase heat transfer. The reasons of this transition in channels below 100 μm diameter will be discussed in Chap. 6.

5.4 Transient Characteristics of the Two-phase Flow

The thermometer resistance fluctuations associated with the phase change have been briefly mentioned in previous chapters. Since the fluctuations were originally recorded from thermometers, they were assumed to be local wall temperature fluctuations due to bubbles forming on and departing from the channel wall. However, more careful examination shows that the amplitudes of the fluctuations correspond to approximately 2-3 °C in temperature, about 10 times bigger than both predictions and observed values in large tubes. A new question is now raised—is this a unique phenomenon in microchannels, or something other than the temperature fluctuation?

5.4.1 Transient Pressure Fluctuations During the Phase Change

Figure 5.10 shows the time traces of the voltage drop across one thermometer with 22 kHz sampling rate before and after the onset of boiling. The thermometer was applied a constant current, and the voltage drop represents the resistance change. The 59 μm diameter channel is one of the devices studied in Chap. 4. With the single-phase liquid flow, the DC voltage signal only carries the white noise; while under two-phase flow, fluctuations with 10-300 Hz frequencies are clearly seen. These patterns of fluctuations are reproducible in repetitive experiments.

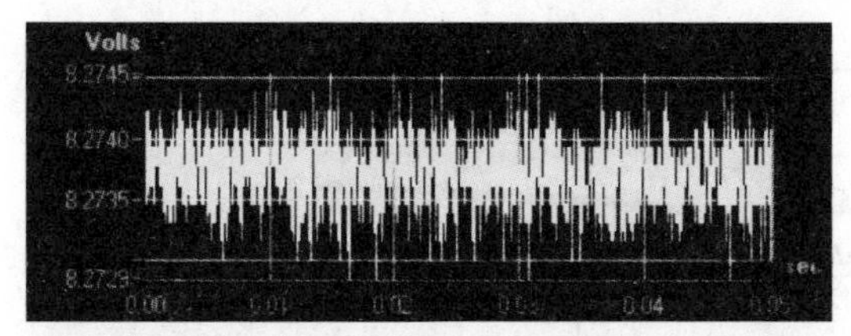

(a) Voltage signal with single-phase flow.

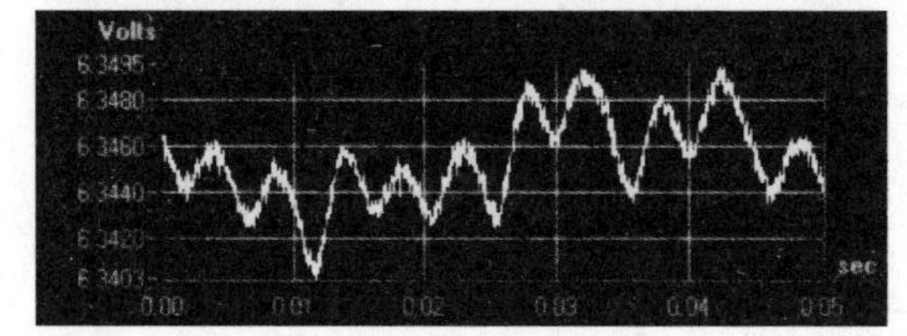

(b) Voltage signal with two-phase flow.

Fig. 5.10 The voltage signals from a thermometer applied with a constant current on a single-channel device, with 59 μm hydraulic diameter. The voltage time trace only shows low-amplitude white noise during the single-phase flow. After boiling begins, there are low frequency fluctuations with approximately 200 Hz frequency.

The automatic data acquisition system introduced in Sect. 3.4 has been used to perform transient measurements. All thermometer resistance signals are simultaneously scanned at 1500 Hz sampling frequency. The average values of every ten samples are then recorded in order to lower the high-frequency white noise, resulting in an effective sampling rate of 150 Hz. Figure 5.11 shows the high-speed time traces of the resistance change of four thermometers along the channel after a constant power is turned on at t=0.5s. The test channel is the 113 μm diameter channel where bubble nucleation has been imaged. The volume flow rate was held constant at 0.1 ml/min, and the effective input heat rate (the amount of heat that is carried by the two-phase fluid) was about 0.8 W. All resistances were sampled for a period of 30 seconds. The nucleation begins at t=11s with slow bubble growth and departure, where all four resistances shows some low-frequency fluctuations. When bubbles depart at higher frequencies along with the development of a boiling flow, the signal frequencies also increase during t=11-18s. With more and more bubbles departing, the annular flow pattern eventually develops during t=18-26s, with an interesting temperature rise at the inlet. And the final steady state two-phase annular flow is developed at t=26s.

Figure 5.12 is an enlarged graph of the time frame t=10-15s in Fig. 5.11, the period from the nucleation to fully developed two-phase flow. Compared with pure liquid flow existing before t=11s, the phase change causes periodic fluctuations at various frequencies. The fluctuation begins with 4-5 Hz frequency, and gradually creeps to 30-40 Hz after t=15s. In the meantime, the amplitude of fluctuation increases at the inlet and decreases at the outlet, indicating nucleation sites are moving towards the inlet.

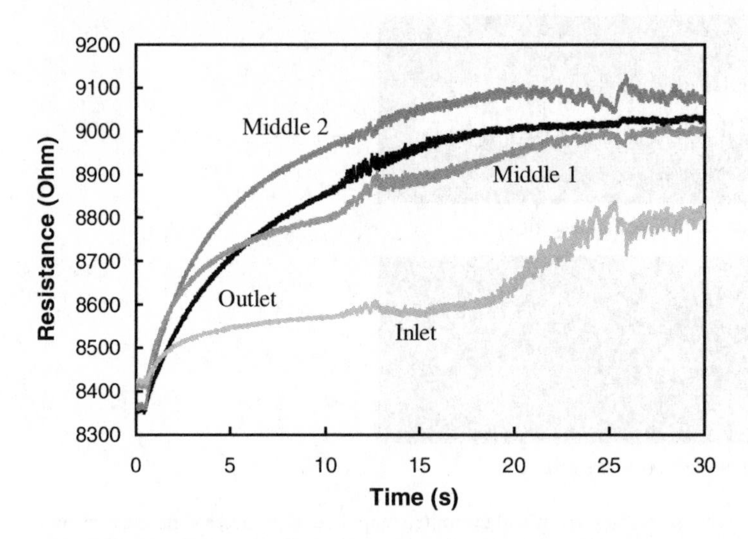

Fig. 5.11 High speed time traces of the thermometer resistance change during the phase change. A constant input heat rate was applied at t=0.5s, and the bubble nucleation was observed at t=11s. The 113 μm diameter channel was tested at 0.1 ml/min constant DI water flow rate.

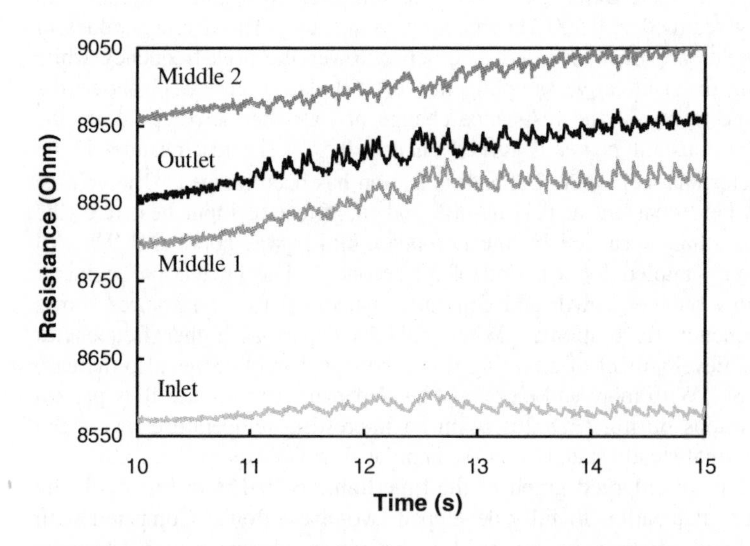

Fig. 5.12 Close-up plot of t=10-15s of the resistance fluctuations in Fig. 5.11.

The recorded transient signals seem to be related to bubble nucleation in the channels. High frequency data acquisition with simultaneous video recording has confirmed this correlation. Figure 5.13 compares the resistance fluctuations of all seven thermometers on one device during the bubble nucleation process. These data were sampled at 150 Hz frequency for 1 second. Although the shape and amplitude of the waveform vary with location, all signals are highly correlated. The videotape recording further confirms that the signal frequency agrees well with the bubble generation frequency. Therefore, the fluctuations are the result of bubble formation and departure from the channel wall.

An important fact is that, the fluctuations in all locations are correlated and independent of the flow pattern at the measurement location. Since the water enters the channel at nearly room temperature, the phase transition always occurs at the middle part of the channel. In other words, the first thermometer at the channel inlet is always under single-phase flow. However, the resistance fluctuation occurs in all locations along the channel, even where the phase change is apparently not happening. This discovery leads to the conclusion that the fluctuations cannot be local wall temperature fluctuations due to a bubble forming and departing on site.

It is well known that doped silicon resistors have both temperature and strain sensitivities. Coincidently, the test channel cross-section dimensions shown in Fig. 5.14 are on the order of conventional pressure sensors; hence the thermometers can have strain response under transient pressure change in the channel. With a pressure calibration unit, which applies a reference pressure to the sealed test channel, the responses of the thermometers were recorded to examine the pressure sensitivity. The pressure calibration of the 113 μm diameter channel shows that the thermometers have varying pressure sensitivities ranging from –1 Ω/psi to –2 Ω/psi depending on the location of the resistor.

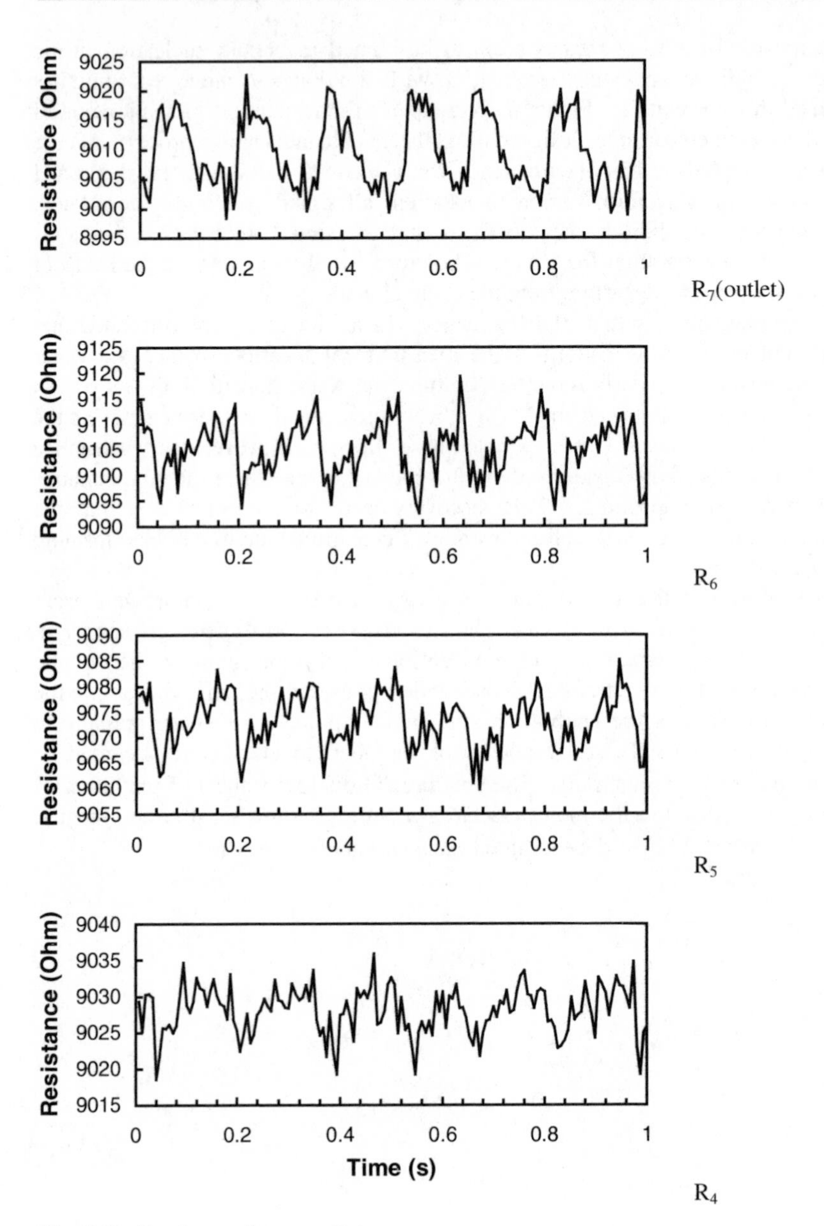

Fig. 5.13 Caption on the opposite page.

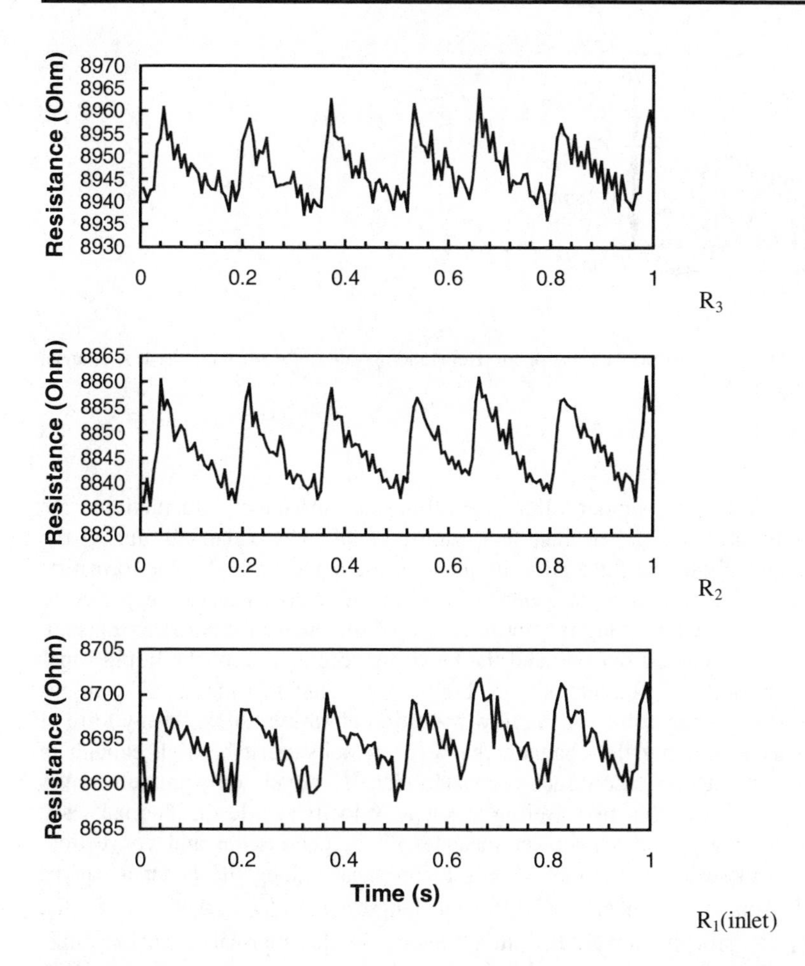

Fig. 5.13 Transient pressure fluctuations at all seven measurement locations along the 113 μm diameter channel. Plots are from the outlet (R_7) to the inlet (R_1). Signals were collected at 150 Hz sampling rate for 1 second.

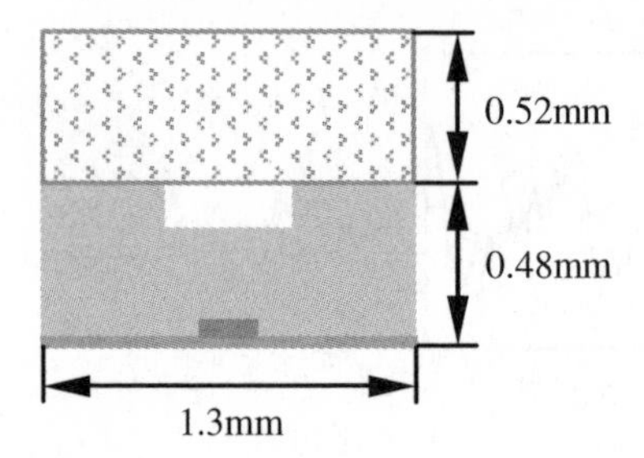

Fig. 5.14 The cross sectional view of the freestanding silicon beam with the test channel and thermometers.

Several evidences support that the transient resistance fluctuations are attributed to pressure rather than temperature in the microchannels during the phase change. First, the fluctuation frequencies are equal to the bubble formation frequencies. The formation of a bubble in a narrow channel can be expected to induce a pressure wave along the entire length of the channel due to the transient increase of the volume flow rate and the resulting acceleration of the liquid-vapor mixture. Second, the fluctuations are highly correlated in time at all locations along the channel regardless of the flow condition at various sites. If any form of wave is traveling along the channel, the wave velocity must be high enough in both water and vapor to allow such correlated signals. Again, only pressure waves can possibly achieve this by traveling at sound velocity, while the thermal wave has to travel through much slower means such as conduction and convection. Third, the timescale for thermal signals to propagate along the channel can be estimated from the concept of diffusion length, $l = \sqrt{\alpha t}$, where α is the diffusivity of silicon (8.921×10^{-5} m^2/s) and t is the thermal diffusion time. Replacing the diffusion length l with the channel length 2 cm, the diffusion time t is found to be approximately 4.5s. Also as shown in Fig. 5.11, the system reaches its steady state approximately 25s after a constant power is applied to the heater. Hence the fluctuations are too fast to be associated with thermal variations at the nucleation sites. Last, assuming the oscillation of liquid-vapor interface is causing localized transient temperature fluctuations, a simple model can be set up for a very small control volume that is filled with saturated liquid and vapor alternatively. This model yields about 0.2 °C of temperature fluctuation, but the measurement result is about 10 times bigger than this value.

Therefore, the resistance fluctuations represent about 10-20 psi transient pressure fluctuations due to bubble formation and departure in the channel. Similar fluctuations have been reported by Kenning and Yan in their 2×1 mm cross section channels [5.4]. These pressure fluctuations are detectable in 113-171 μm diameter channels where bubble nucleation can be clearly seen. But no apparent fluctuation has been detected in 27-44 μm diameter channels. Part of the reason is that the eruption boiling does not have a measurable bubble growth time. In addition, the pressure sensitivity of the thermometers is much smaller due to thicker channel walls. However, the expected transient pressure fluctuation in smaller channels is even higher due to higher evaporation rate. The transient pressure wave also explains why some devices can be damaged during phase change experiments.

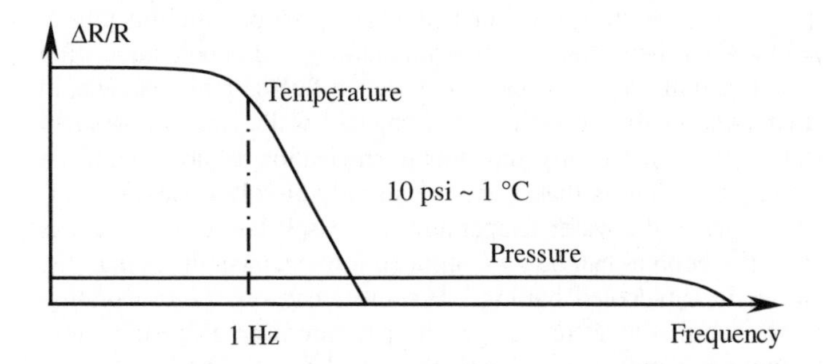

Fig. 5.15 Comparison of frequency response curves of the temperature and pressure measurement with the same doped silicon resistor.

Finally, the question of how reliable the measurements are has to be answered, given that the same sensor has both temperature and pressure sensitivities. As shown in Fig. 5.15, the two sensitivities have different frequency responses. When the resistors are used as thermometers, temperature measurements are performed in DC (Direct Current) and very low frequency regions, where the pressure sensitivity is insignificant compared with the temperature sensitivity. With 10 psi to 1 °C equivalence in the two domains, the transient fluctuations only cause about 1-2 °C uncertainty in the steady state temperature measurements. The temperature sensitivity decreases quickly for signals above 1 Hz due to the necessary thermal diffusion time in the system. For the transient pressure, the signals of interest are in the greater than 1 Hz frequency region. Therefore, the AC (Alternating Current) signals carried by the DC temperatures are considered pressure components.

5.4.2 Pressure Drop During the Phase Change

A pressure transducer is placed at the entrance of the fixture to measure the global pressure drop along the microchannel. The fixture has internal flow channels that are much larger than the microchannels. Due to the damping of the transient pressure fluctuations in these flow channels, the transient signals cannot be detected by the pressure transducer. However, the phase change process can still be judged in the pressure drop along the entire channel. Figure 5.16 is the simultaneous pressure drop measurement with the resistance measurements shown in Fig. 5.11. The pressure signal was continuously sampled at 150 Hz. As seen in all phase change pressure curves, the macroscopic pressure drop along the channel decreases in the single-phase flow regime due to the decrease in the liquid viscosity, and immediately increases after the phase change due to the sudden increase in the volume flow rate.

However, the pressure curve clearly indicates that the onset of boiling begins at t=8s, instead of t=11s when the initial pressure fluctuations and bubble nucleation are detected. At t=11s, only a very small increase or fluctuation is shown in Fig. 5.16. Apparently the bubble growth on the channel walls causes this small pressure fluctuation. The reason why the bubble nucleation is not always an indication of the onset of boiling is that, there is naturally dissolved gas in water; and the gas is released when the water temperature increases due to its lower gas solubility. Therefore the bubbles can be seen at much lower temperatures than the boiling point of water in sub-cooled boiling. However, because the evaporation rate is low and is not continuous at this stage, the pressure drop along the entire channel does not show an overall increase until the actual onset of boiling begins. The time frame between the nucleation and onset of boiling, about 7s in the figure, agrees well with the time before the bubbles turn to continuous annular flow on the video recordings.

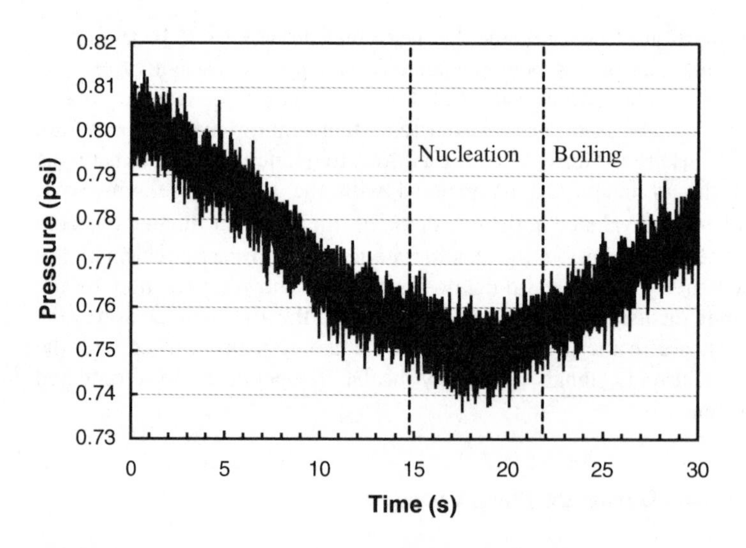

Fig. 5.16 Pressure drop along the channel during the phase change. A constant power was applied at t=0.5s and the pressure change was sampled at 150 Hz for 30 seconds. The measurement was performed simultaneously with the resistance measurements in Fig. 5.11. The nucleation begins at around t=11s and the onset of boiling occurs at t=18s.

5.4.3 Transient Wall Superheat During the Phase Change

Temporal wall superheat can occur even after the onset of boiling due to local bubble formations. As discussed in the two-phase microchannel heat sinks in Chap. 2, a local temperature drop of about 5 °C was measured at the outlet when the first bubble formed in one of the microchannels. Also as shown in Fig. 5.17, a transient wall temperature superheat of 5-8 °C is recorded in a period of approximately 30s in a 61 μm diameter channel. This transient superheat explains the wall temperature reduction after the onset of boiling shown in Fig. 4.6.

The temporal superheat is attributed to the conditions of active nucleation sites within the microchannels. Due to the small channel wall area and relatively low surface roughness in plasma etched channels, the number of active nucleation sites in the microchannels can be very limited and as a result, the condition change at only a few nucleation sites can significantly affect the phase change in the entire channel.

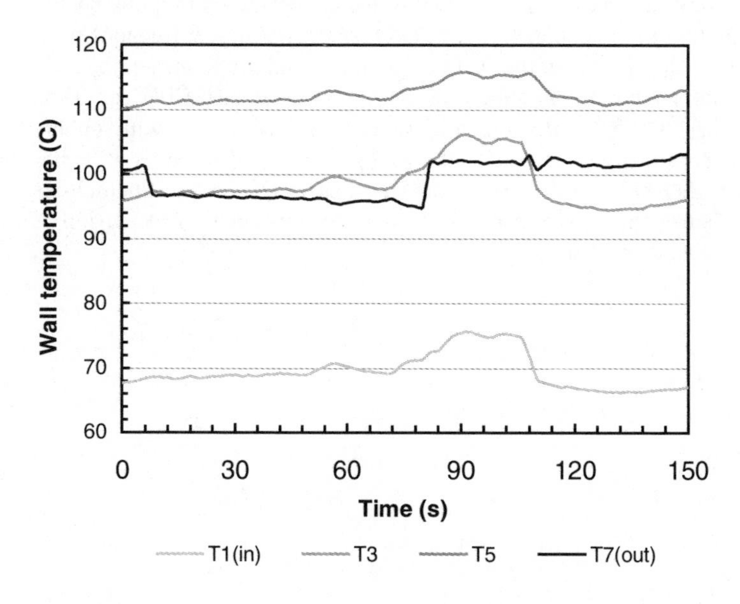

Fig. 5.17 Transient wall superheat during the phase change in a 74 μm wide, 52 μm deep, 61 μm diameter channel at 0.1 ml/min DI water flow rate. The wall temperature changes at four locations were continuously sampled every 2 seconds for a period of 150 seconds.

5.5 Discussion

The bubble nucleation and the resulting transient pressure fluctuations have been discussed in this chapter. Bubble formation in plasma etched silicon microchannels has been confirmed with optical visualizations. Simultaneous optical high-speed recording and transient state measurements verify that bubbles can form and grow in sub-150 μm diameter microchannels, and cause large transient pressure fluctuations upon departure in the channel.

Experiments show that there is a boiling mechanism transition in plasma etched silicon microchannels with plain walls below 100 μm diameter. In channels larger than 100 μm, the phase change begins with small bubble growth on the side walls. The bubbles continue to grow after departing the channel wall and eventually induce wavy annular flows in the steady state. While in channels smaller than 50 μm, the phase change occurs suddenly at a certain input heat rate, and the liquid-vapor mixture quickly forms after a very short annular flow region. The reasons have been found to be smaller nucleation sites in smaller channels, which will be discussed in the next chapter.

Both transient temperature and pressure fluctuations can occur in microchannels due to the bubble formation. The transient temperature fluctuation results from local wall superheat during bubble nucleation and usually has periods of longer than 1 second. The transient pressure fluctuation results from the bubble growth and departure in the channel, and usually happens at higher frequencies of greater than 3 Hz. Depending on the surface roughness and active nucleation sites in the channel, the temporal wall superheat before the onset of boiling can be as high as more than 10 °C. The pressure fluctuations associated with bubble formation and departure are on the order of 10-20 psi and could be larger at higher flow rates and higher heat fluxes. Since the transient large pressure fluctuations can potentially damage the device package, it must be considered in the design of two-phase microchannel heat sinks.

6 Enhanced Nucleate Boiling in Microchannels

6.1 Background: Bubble Nucleation in Microchannels

In Chap. 5, two boiling mechanisms, nucleation and eruption boiling, have been visualized in microchannels. For channel diameters ranging from 100 to 150 μm, boiling initiates from bubble nucleation on the side walls and gradually turns into annular flows in the fully developed state; for channels smaller than 50 μm, on the other hand, boiling initiates from a sudden eruption of vapor at much higher temperatures and mist flow forms immediately. The eruption style boiling has also been observed in some previous research. It has been assumed to be due to the size effect, and hypotheses such as "evaporating space" have been proposed to explain this unusual boiling mechanism and the associated large amounts of wall superheat. However, this theory has not been experimentally confirmed because the reported critical size or space is inconsistent from various experiments, with a range from 50 to 700 μm [5.2-5.5]. In this chapter, a new theory is proposed and experimentally examined, a theory that relates the boiling mechanisms to the active nucleation sites in the channel walls.

6.1.1 "Evaporating Space" Hypothesis

Peng et al. proposed one of the earliest hypotheses to explain how the channel dimension affects the boiling mechanism [6.1-6.2]. They suggest that there exists an "evaporating space" and if the channel size is smaller than this threshold, eruption boiling with typical large amounts of wall superheat will be induced.

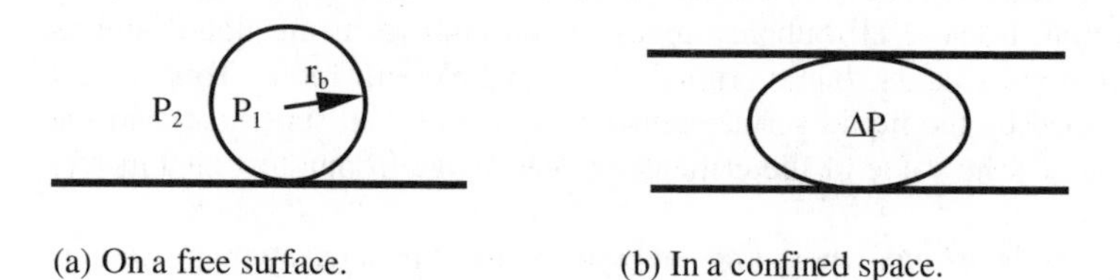

(a) On a free surface. (b) In a confined space.

Fig. 6.1 Mechanisms of the vapor bubble growth in different spaces.

Figure 6.1(a) shows a vapor bubble growing on a free surface as in the case of pool boiling or boiling in large channels where the bubble diameter is negligible compared to the channel dimension. In order for the bubble to remain in the liquid and keep growing, the internal pressure P_1 must overcome the ambient liquid pressure P_2 and the surface tension σ at the liquid-vapor interface:

$$\Delta P = P_1 - P_2 \geq \frac{2\sigma}{r_b} \tag{6.1}$$

where r_b is the radius of the bubble.

In the case of bubble nucleation in a very small channel as shown in Fig. 6.1(b), the bubble is forced to grow laterally; hence the required pressure difference ΔP^* is expected to be larger than ΔP. Peng et al. further proposed a correction factor as a function of the radius of the bubble r_b and the hydraulic diameter of the channel D_H, so that

$$\Delta P^* = \Delta P \cdot f(2r_b / D_H) \tag{6.2}$$

Substituting Equation (6.2) and the Clausius-Clapeyron equation into the nucleation condition, the degree of required wall superheat ΔT_{sup} is determined by

$$\Delta T_{sup} = T_w - T_{sat} \geq \frac{4CT_{sat}(v_v - v_l)\sigma}{h_{fg} D_H} \tag{6.3}$$

where C is an empirical constant, T_{sat} is the liquid saturation temperature, v_v and v_l are the specific volumes of the vapor and liquid at saturation temperature respectively, and h_{fg} is the latent heat. Equation (6.3) indicates that the degree of wall superheat is a function of the size of the channel at a given surface tension σ. The authors further suggest that this critical channel diameter is on the order of 100 μm.

6.1.2 Heterogeneous Nucleation on a Solid Surface

Another factor that possibly affects the nucleation process is the wall surface condition. In the fundamental boiling theories, the phase change consists of two processes—bubble formation and bubble growth. The bubble formation is also called nucleation, because all bubbles appear as an embryo in the liquid and as long as the embryo is larger than a critical size, a bubble will form. This critical size is determined by the liquid surface tension, degree of wall superheat, and the ambient pressure. The value of the critical size can range from a few nanometers to micrometers.

Depending on how and where the embryo forms, the nucleation process is divided into homogeneous nucleation and heterogeneous nucleation [6.3-6.4]. If the vapor embryo completely forms within a superheated liquid, it is called "homogeneous nucleation". This form of nucleation is usually associated with high degrees of wall superheat, up to the critical temperature of the liquid (370 °C for water), and extremely rapid rates of vaporization. The vapor may or may not form a visible bubble, and even vapor explosions may occur in some instances. On the other hand, if the vapor embryo forms at an interface between a superheated liquid and another phase that the liquid contacts, such as a solid or gas, it is called "heterogeneous nucleation". The so-called nucleate boiling actually refers to heterogeneous nucleation, with visible bubble growth on a solid surface at low superheat. When heterogeneous nucleation occurs on an ideally smooth surface, the nucleation process takes the same form as homogeneous nucleation. However, due to the presence of trapped gas in crevices and scratches on a solid surface in reality, heterogeneous nucleation always happens at much lower temperatures than the critical temperature of the liquid.

For heterogeneous nucleation from a cavity on a surface, the Clausius-Clapeyron equation for superheat is used together with the Gaussian expression to relate the wall superheat ΔT_{sup} to the surface tension σ and the radius of the cavity r_c [6.4]:

$$\Delta T_{\sup} = \frac{2\sigma T_{sat} v_v}{h_{fg} r_c} \tag{6.4}$$

However, not every cavity becomes active during the phase change because the nucleation process is dictated by multiple factors. For a given amount of imposed wall superheat, the minimum cavity radius, referred to as the critical cavity size r^*, is given by

$$r^* = \frac{2\sigma T_{sat} v_v}{h_{fg} \Delta T_{\sup}} \tag{6.5}$$

Hsu studied bubble nucleation in pool boiling and proposed models that considered the thermal boundary layer thickness [6.5]. The thermal boundary layer affects the nucleation process because a finite heating time is required in order for a bubble to form in a cavity. This also explains why a certain amount of wall superheat is always required in nucleate boiling. Hsu's model takes the heating time into consideration and suggests that only the cavities with a finite waiting period can become active nucleation sites. The range of the opening radius of the active cavities is given by:

$$r_{c,\max}, r_{c,\min} = \frac{\delta}{2C_1}\left[1 - \frac{T_{sat} - T_\infty}{T_w - T_\infty} \pm \sqrt{\left(1 - \frac{T_{sat} - T_\infty}{T_w - T_\infty}\right)^2 - \frac{8\sigma T_{sat} v_v C_3}{h_{fg}\,\delta\Delta(T_w - T_\infty)}}\right] \tag{6.6}$$

where δ is the thermal boundary layer thickness, T_w is the wall temperature, T_∞ is the bulk liquid temperature, C_1 and C_3 are constants related to the shape of the cavity and the contact angle of the bubble embryo. For simplicity, Hsu assumed the bubble embryo had the same diameter as the cavity mouth, and obtained the values as 2 and 1.6 respectively.

The "evaporating space" hypothesis explains the eruption boiling at unusual superheat with the liquid surface tension; while the heterogeneous nucleation theories indicate that the lack of active nucleation sites also leads to large amounts of wall superheat. It is known that adding cavities and trenches on a macroscale boiling surface, known as surface enhancement, helps initiate nucleate boiling with low superheat, and the technology has been used in the industry for more than twenty years [6.6-6.9]. More recently, some research shows that homogeneous nucleation could happen on smooth microscale surfaces. Klausner and Mei demonstrated that the nucleation site density was correlated with the ratio of the maximum to minimum cavity radius, and they suggested that complete suppression of nucleation sites occurred for the ratios ranging from 50 to 150 μm [6.10]. Lin and Pisano observed bubble formation on a surface micromachined, smooth line heater at temperatures close to the superheat limits of three dielectric liquids at atmospheric pressure, which represents the case of heterogeneous nucleation on a smooth surface [6.11]. Kwat et al. did extensive modeling on homogeneous bubble growth [6.12-6.14]. In this chapter, channels with various surface roughness and liquids with various surface tensions are studied and the results show that the number of active nucleation sites takes a leading role in determining the nucleation mechanism in microchannels below 100 μm diameter.

6.2 Enhanced-wall Microchannel Test Devices

In all experiments presented in earlier chapters, only microchannels formed by DRIE with plain side walls have been used. This chapter focuses on how the liquid surface tension and wall surface roughness affect the boiling mechanism. Some of earlier experiments have been repeated with liquids that have varying surface tensions to examine how the surface tension affects the nucleate boiling. Also by varying the DRIE parameters, microchannels with different surface roughness were obtained. However, micromachining only provides very limited change in the surface roughness. In order to further study how gas-trapping cavities help the nucleate boiling in microchannels, test devices with enhanced walls were also used.

6.2.1 Nucleation Sites in Plain-wall Channels

Although the visualizations of bubble nucleation in Chap. 5 does not directly answer the question how the surface condition affects the nucleation process, one interesting fact is that all recorded nucleation sites are on the side walls rather than on the bottom of the channel. The SEM images in Fig. 6.2 answer the question why the bubbles prefer to grow from the side walls. The alternating etch and passivation cycles in the DRIE process form scallops of 0.2-0.4 μm and some larger defects on the side wall, hence the surface roughness on the side walls is orders of magnitude higher than that of the bottom. This is the first evidence that the surface conditions may play a role in the microscale nucleation process.

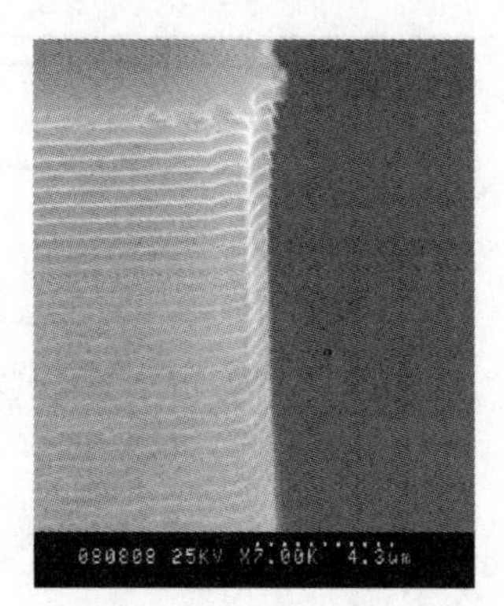

Fig. 6.2 SEM images of the channel side walls with standard etch parameters. Due to the etch/passivation cycles during plasma etch, 0.2-0.4 μm scallops are formed on the side walls.

6.2.2 Design Parameters of Test Devices

The predicted active nucleation site dimensions as a function of wall superheat is shown in Fig. 6.3, using both critical cavity size r^* from Equation (6.5) and Hsu's model in Equation (6.6). Since Hsu's model was developed for pool boiling, some assumptions were applied for the simulation in microchannels as follows. The thermal boundary layer thickness is assumed to be a half of the channel diameter. That is, the simulated 10-150 μm thick thermal boundary layers correspond to 20-300 μm diameter channels. The bulk liquid temperature T_∞ is the sub-cooling temperature in the original model for pool boiling. However, in the forced internal flow, due to the varying liquid temperature profile from the inlet to the outlet, local T_∞ is defined as the average liquid temperature in the specific cross section. For simplicity, saturated boiling is assumed, that is, T_∞ is equal to 100 °C (T_{sat} at 1 atm). The minimum cavity sizes from both models are in good agreement. In addition, since Hsu's model considers a finite heating time, there is a maximum cavity site limit as well as minimum wall superheat for a certain boundary layer thickness. The predictions indicate that for 20-300 μm diameter channels, cavities with 2-10 μm opening mouth help limit the wall superheat within 5 °C.

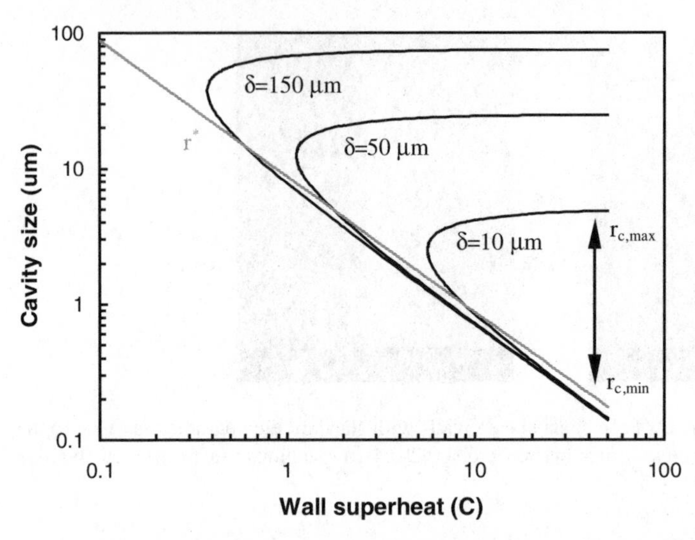

Fig. 6.3 Predictions of active nucleation site sizes as a function of wall superheat and boundary layer thickness. The critical cavity size r* predicts the lower bound of the cavity size as a function of wall superheat. Hsu's model further considers the finite heating time due to limited thermal boundary layer thickness, hence giving an upper bound of the cavity size as well. The minimum sizes are in good agreement in the two models.

Table 6.1 Structural parameters of enhanced-wall channels.

Structure	Dimension
Reservoirs	600 μm wide, 250 μm deep, 1 cm long
Freestanding beam	1.3 mm wide, 2 cm long
Channel dimensions	10-150 μm wide, 10-200 μm deep, 2 cm long
Aluminum heater	50 μm wide, 10×2 cm long, 2 μm thick resistance 60 Ω
Resistive thermometers	150 μm long, 50 μm wide, 1 μm junction depth resistance 8 kΩ
Side wall notches	4 μm wide opening, 20 μm deep
Bottom cavities	4-8 μm diameter, 150-250 μm deep
Si substrate thickness	480 μm
Glass thickness	520 μm

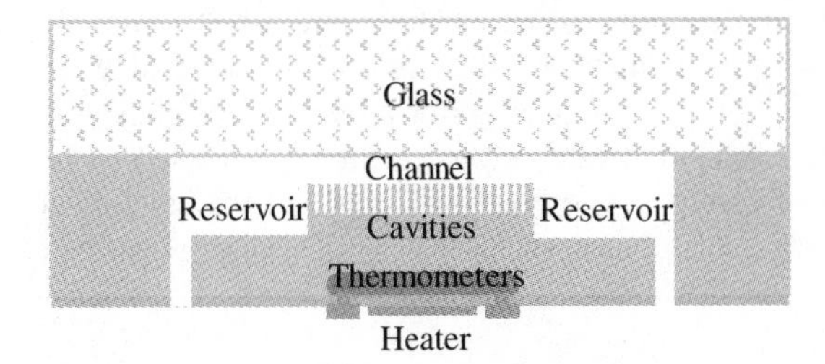

Fig. 6.4 Cross sectional view of the test channel with an aluminum heater and seven doped silicon thermometers. Channels were fabricated with various surface conditions by changing the etch parameters or making notches and cavities in the channel walls.

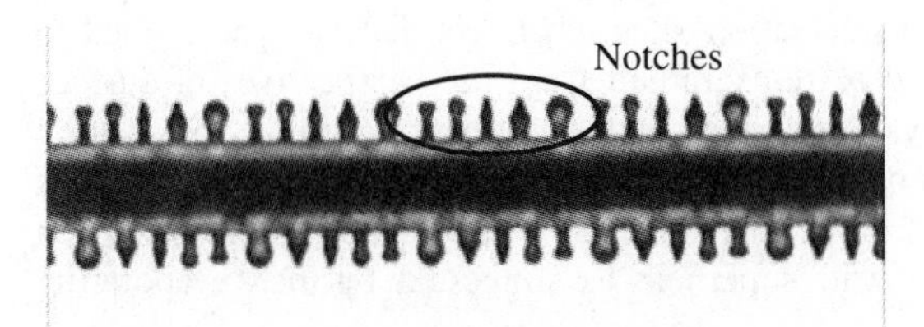

(a) Notches with 4 μm opening, 20 μm depth into the side walls.

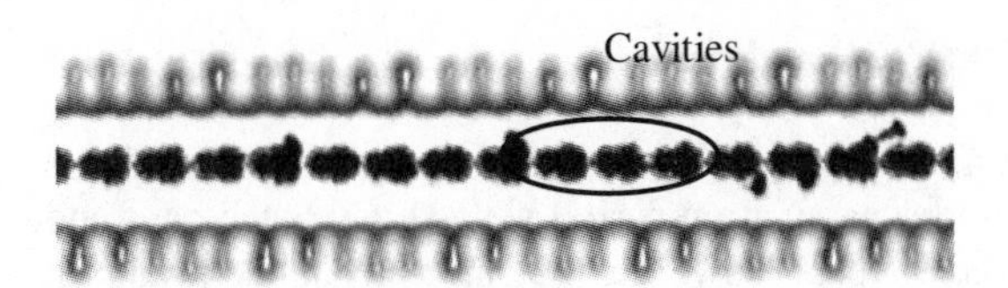

(b) 4-8 μm diameter, 150-250 μm deep cavities in the bottom of the channel.

Fig. 6.5 The notches in the side walls and the cavities in the bottom as surface enhancement. They are designed to trap air to facilitate nucleation process during the phase change.

The instrumented microchannels have an aluminum heater and seven doped silicon thermometers integrated and were configured for automatic data acquisition. Different DRIE programs were used to obtain various surface roughness. 4-8 μm notches and cavities were made into selected channels with diameters below 100 μm. The design parameters are listed in Table 6.1.

Figure 6.4 is the cross sectional view of the glass-sealed silicon channel with enhanced wall surfaces. Channels with notches and cavities in the walls are shown in Fig. 6.5. These added features were designed to trap air in the channels and become active nucleation sites as predicted from heterogeneous nucleation theories. The notches have 4 μm openings to the channel, and are 20 μm deep into the channel wall. The 4-8 μm diameter cavities in the bottom wall were formed at the same time with the 250 μm reservoirs prior to etching the channels. Because of the small area, the DRIE etch rate of the cavities is expected to be slower than that of the larger reservoirs. The estimated depth of the cavities is 150-250 μm.

6.3 Phase Change in Silicon Channels with Plain Walls

Microchannels with 27-171 μm hydraulic diameters and varying surface roughness were used to carry out experiments with water that had varying surface tensions. Because the surface condition can play an important role in gas trapping and determining the number of nucleation sites, channels that have been used to make transient measurements were re-examined from the perspectives of surface roughness and the degree of wall superheat at the onset of boiling. The general conclusions from the experiments are: the DRIE induced surface roughness does not significantly affect the boiling mechanism; and a lower liquid surface tension does not apparently lower the wall superheat as suggested by the "evaporating space" hypothesis.

6.3.1 Plasma Etch Induced Surface Roughness

Table 6.2 lists the etch parameters used in various etch programs. In all the parameters, longer etch with shorter passivation period yields higher etch rates and angled side walls due to isotropic etch. However, the surface roughness falls in the same order of magnitude with all etch programs. In other words, plasma etched silicon surfaces achieve less than 1 μm surface roughness.

Table 6.2 STS DRIE system program parameters and resulting surface roughness.

Etch parameters	Standard	Etch 1	Etch 2	Etch 3
Etch time (s)	12	5	24	7
Passivation time (s)	7	5	3	5
Base pressure (mTorr)	1	1	1	1
Pressure trip (mTorr)	94	94	94	94
SF6 flow (Sccm)	130	130	130	130
C4F8 flow (Sccm)	85	120	85	85
RF power (W)	120	120	120	120
Coil power (W)	600	600	600	600
Etch rate (μm/min)	3.25	1.67	4.50	2.67
Side wall roughness (μm)	0.2-0.4	0.2	<0.1	0.3
Bottom wall roughness (μm)	<0.1	0.4*	0.9	<0.1

* *with 1-15 μm tall bumps.*

6.3.2 Plasma Etched Silicon Channels with DI Water

The boiling mechanisms and two-phase flow regimes in 27-171 μm diameter channels have been discussed in Chap. 5, and boiling images in a 113 μm diameter channel and a 44 μm diameter channel were shown. The phase change in these two channels is representative for channels larger than 100 μm and channels smaller than 50 μm respectively. The same two channels were used in the wall temperature profile measurements again to examine the wall superheat associated with each boiling mechanism. In order to study whether and how surface roughness affects the boiling mechanism, two channels of similar dimensions were examined with different surface roughness.

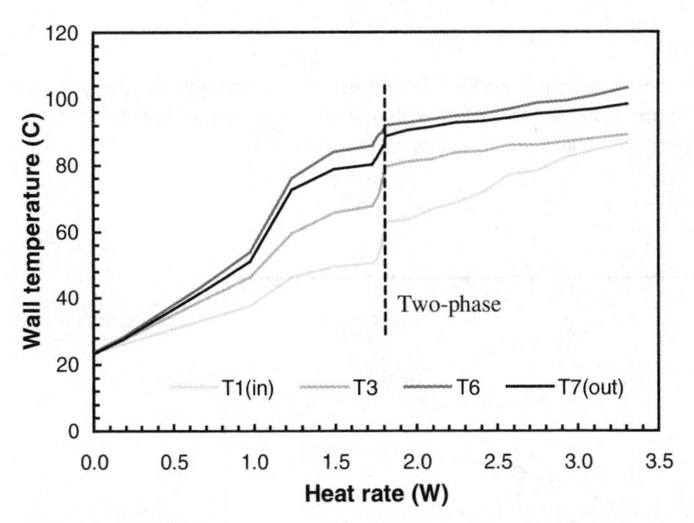

Fig. 6.6 Wall temperature change from four thermometers as a function of input heat rate in a 100 μm wide, 130 μm deep, 113 μm hydraulic diameter channel at 0.1 ml/min DI water flow rate. The thermometers have ±0.5 °C temperature accuracy.

Figure 6.6 shows the wall temperature measurement from four thermometers on the 113 μm diameter channel with increasing heat rate. The previous phase change visualization shows that the boiling initiates from small bubble growth in the channel, and the annular flow regime is the dominant flow pattern. The initial bubble nucleation induced by out-gassing in the liquid, also called "partial boiling", causes a small temperature plateau in the boiling curve because the bubble departure generates intermittent annular flows in the channel, which increase the local heat transfer coefficient, but not as high as in the two-phase region. When the boiling begins, the two-phase annular flow dominates in the channel, yielding a very high heat transfer coefficient; hence a second temperature

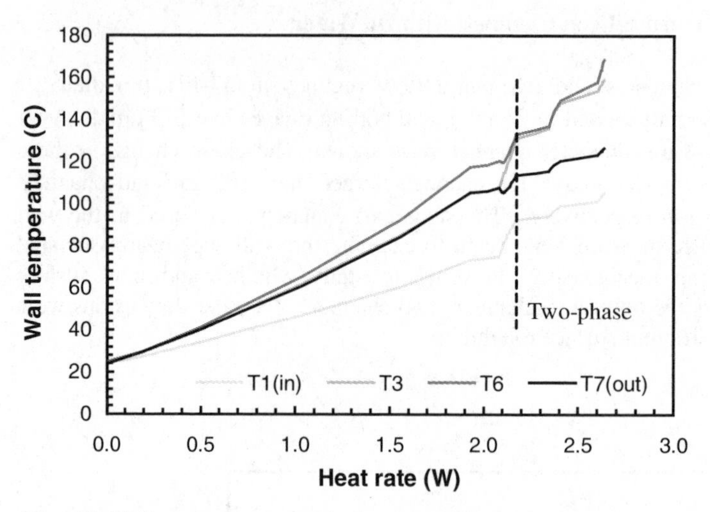

Fig. 6.7 Wall temperature change from four thermometers as a function of input heat rate in a 50 μm wide, 40 μm deep, 44 μm hydraulic diameter channel at 0.02 ml/min DI water flow rate. The thermometers have ±0.5 °C temperature accuracy.

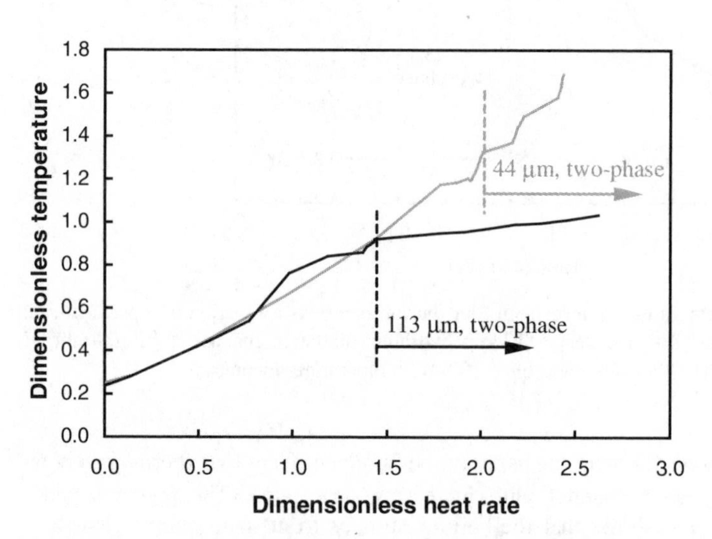

Fig. 6.8 Comparison of the boiling curves from normal nucleate boiling and eruption boiling. Two local wall temperature curves (T6) from Figs. 6.6 and 6.7 are plotted in the dimensionless form. Due to the eruption boiling, the phase change in the 44 μm channel is "postponed" to a higher heat rate and wall temperature.

plateau results. Partial boiling begins at approximately 80 °C as the gas bubbles form due to lower gas solubility at higher liquid temperatures, and they turn to vapor cores later when the temperature reaches the boiling point. The measured wall superheat after the onset of boiling is below 5 °C.

Figure 6.7 shows the boiling curves for the 44 μm diameter channel. Just as the clear difference in the boiling mechanisms, the boiling curves in the two typical channels are also different. The 44 μm channel shows an eruption of vapor with more than 20 °C wall superheat at the onset of boiling. There is no clear temperature plateau in the two-phase region, indicating that mist flow or even dry-out spot forms immediately after the phase change. Because the mist flow has much lower heat transfer coefficients than those of the annular flow, the wall temperatures keep rising with increasing heat rate.

From the thermal circuit introduced in Sect. 3.2, the percentage of the input heat rate that was carried by the fluid in the two experiments was found to be 42% for the 113 μm channel and 10% for the 44 μm channel respectively. Figure 6.8 compares the highest wall temperature spot (*T6*) from the two channels, in the dimensionless form with normalized temperature and heat rate from Equation (2.6). The wall temperature is divided by the saturation temperature at the measured pressure drop, and the heat rate is divided by the sensible heat of water at the fixed flow rate. In the 113 μm channel, the thermal resistance is approximately 95.2 °C/W in the single-phase region and 15.6 °C/W in the fully developed two-phase region. The measurement shows an 84% reduction in the thermal resistance because of the phase change. With a long temperature plateau and low superheat in the two-phase region, the boiling curve from the 113 μm channel is ideal for microchannel heat sinks. In the same plot, however, the phase change in the 44 μm channel seems to be "postponed" to a higher heat rate and wall temperature. Due to the eruption boiling, the heat transfer does not seem to be more efficient after the phase change, with approximately 600 °C/W in both single- and two-phase regions.

Two more channels have been fabricated using etch programs No. 1 and 2 in Table 6.2, with 0.4 μm and 0.9 μm surface roughness respectively. Both channels have hydraulic diameters between 50-70 μm, which were designed to examine whether the increased surface roughness would induce bubble nucleation as in channels larger than 100 μm. The experiments showed more than 30 °C wall superheat in both channels. Figure 6.9 shows the wall temperature change against the input heat rate in a 50 μm wide, 70 μm deep channel. Phase change did not occur even at 145 °C wall temperature, while the boiling point of water was 103.6 °C at the measured pressure drop along the channel. The same amount of superheat still existed at reduced flow rates.

In summary, channels with 27-171 μm hydraulic diameters have been examined, and all channels smaller than 50 μm show an eruption of vapor with 20-50 °C wall superheat rather than bubble growth at below the boiling point as in larger channels. These experiments also indicate that the surface roughness of less than 1 μm after plasma etch dose not apparently lower the superheat required for

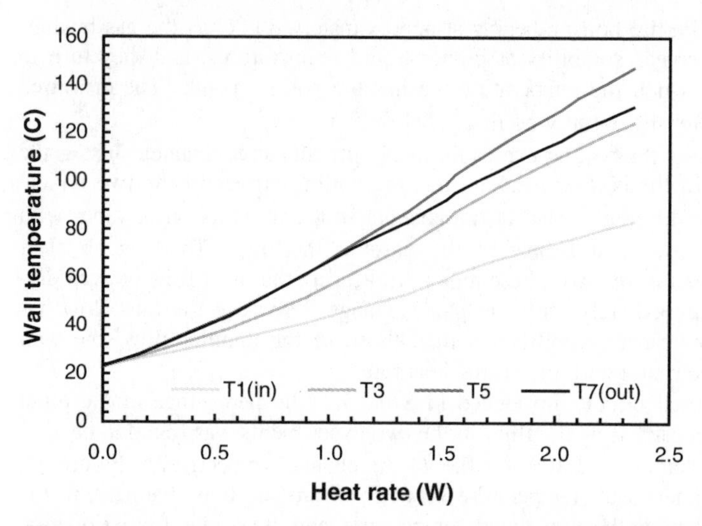

Fig. 6.9 Wall temperature change from four thermometers as a function of input heat rate in a 50 μm wide, 70 μm deep, 58.3 μm hydraulic diameter channel at 0.075 ml/min DI water flow rate. Boiling was not observed in the measurement. The thermometers have ±0.5 °C temperature accuracy.

nucleation. This result is consistent with Equation (6.4), which predicts 18-88 °C wall superheat with 0.2-0.9 μm cavity opening diameters. The bubble nucleation in larger than 100 μm channels is very likely formed in larger defects that do not exist in smaller channels.

6.3.3 Plasma Etched Silicon Channels with Surfactant

From Equation (6.3), the "evaporating space" hypothesis suggests that the wall superheat is a linear function of the liquid surface tension given a certain channel dimension. In consequence, when different liquids are used in the same heated microchannel, the low-surface tension liquid is expected to boil with lower degrees of wall superheat. Experiments have been designed to verify this prediction.

Reduction of DI Water Surface Tension

The surface tension of DI water can be lowered by adding a surfactant, Triton X-100. The contact angle of the liquid-gas phase on a solid surface is a good measurement of the liquid wetting properties on the surface and the surface tension. As shown in Fig. 6.10, the contact angles of the liquid droplets decrease with increasing surfactant concentrations. With these solutions that have reduced surface tensions, the wall superheat is expected to be lower from "evaporating space" hypothesis.

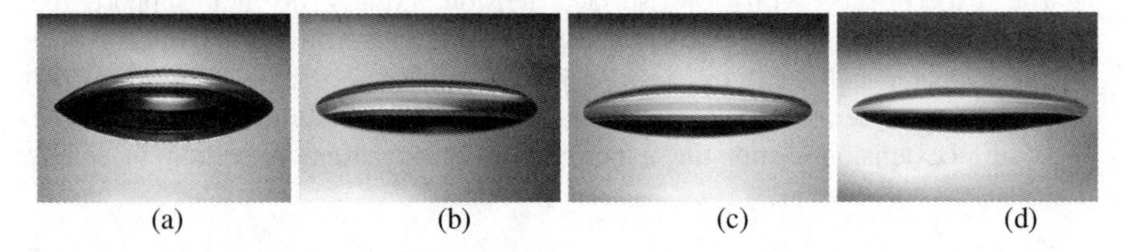

Fig. 6.10 Wetting properties of water on a polished silicon surface. (a) DI water. (b) Triton X-100 solution, 200 ppm (weight concentration). (c) Triton X-100 solution, 2000 ppm. (d) Triton X-100 solution, 10000 ppm.

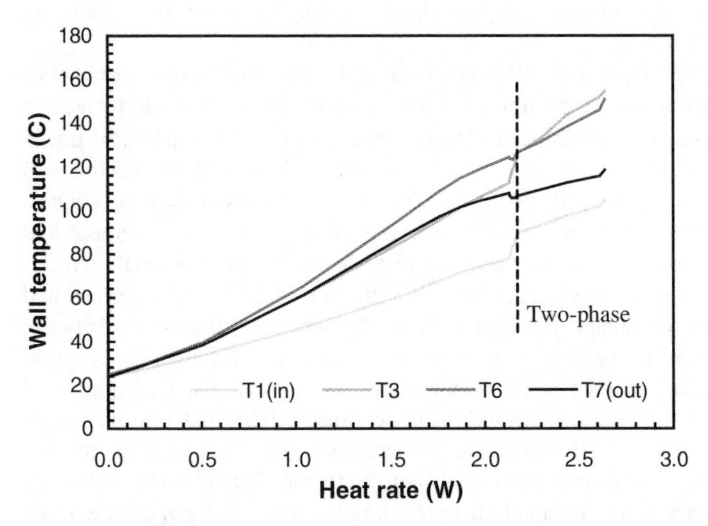

Fig. 6.11 Wall temperature change from four thermometers as a function of heat rate in a 50 μm wide, 40 μm deep, 44 μm hydraulic diameter channel at 0.02 ml/min flow rate of 200 ppm TritonX-100 DI water solution. The thermometers have ±0.5 °C temperature accuracy.

Wall Temperature Measurement During Phase Change

All four liquids shown in Fig. 6.10 have been used in the 44 μm channel in the wall temperature superheat experiment. The result of DI water was shown in Fig. 6.6, and Fig. 6.11 is the wall temperature measurement with 200 ppm Triton X-100 solution. The eruption boiling mechanism was observed again in this experiment. Further experiments with 2000 ppm and 10000 ppm solutions showed similar boiling curves as in Fig. 6.11, within 4 °C temperature difference. Therefore, the nearly 20 °C wall superheat is not directly related to a high surface tension.

The experiments with low surface tension liquids do not support the "evaporating space" hypothesis. In addition, bubble growth sizes of smaller than 10 μm diameter and departure sizes of 50-100 μm diameter have been visualized in earlier experiments with larger channels. These observations indicate that the high surface tension is not the major reason of the eruption boiling in small-diameter microchannels.

6.3.4 Boiling Regime Chart for Plain-wall Channels

Summarized boiling mechanisms and two-phase flow regimes in microchannels with 27-171 μm diameters are given in Table 6.3. The boiling mechanisms in channels with various surface roughness and various DI water flow rates are compared.

According to the facts and observations in the table, the boiling mechanism transition happens at around 50 μm for plasma etched silicon channels with plain side walls. Figure 6.12 plots phase change chart for channels below 150 μm in hydraulic diameter, including the dominant two-phase flow regimes and the wall superheat before the onset of boiling. As the experimental results suggest, channels larger than 100 μm usually have less than 5 °C wall superheat and annular flow develops as the result of nucleate boiling; while channels smaller than 50 μm have vapor eruption mechanism with more than 20 °C superheat and mist flow is the dominant two-phase flow regime; the dimensions between 50-100 μm show either nucleation or eruption mechanism. Although the eruption boiling carries some characters of homogeneous nucleation, the measured wall superheat still falls in heterogeneous nucleation mechanism and the amount of superheat is consistent with theoretical predictions. This fact implies that the bubble nucleation still initiates from small defects in the channel wall. However, due to the geometry limits in small channels, large defects (2-5 μm cavities) that require lower superheat do no exist. When there is not enough gas trapped in the channel walls, the nucleation mechanism shifts more to the homogeneous side, requiring larger amounts of wall superheat.

Table 6.3 Boiling regime summary in 27-171 μm diameter microchannels with plain walls.

Hydraulic diameter (μm)	Channel width (μm)	Channel depth (μm)	Flow velocity (m/s)	Phase change mechanism/ flow regime	Wall superheat (°C)
26.7	20	40	0.063	Eruption/mist	20-30
44	50	40	0.125, 0.167, 0.250, 0.417	Eruption/mist	10-20
45.7*	50	42	0.079	Eruption/mist	20-50
52	74	40	0.113	Nucleation/ annular or eruption/mist	0-5
58.3**	50	75	0.238, 0.357	Eruption/mist	>50
60	120	40	0.174, 0.347	Nucleation or eruption/mist	0-5
72	74	70	0.032, 0.161, 0.322	Eruption/mist	0-10
88.4	120	70	0.198, 0.496	Nucleation/mist	0-20
97.4	100	95	0.035, 0.175, 0.439	Nucleation/ annular	0-5
113	100	130	0.026, 0.128, 0.321, 0.641	Nucleation/ annular	0-5
138	250	95	0.070, 0.175	Nucleation/ annular	0-5
171	250	130	0.256, 0.384	Nucleation/ annular	0-5

* *with 0.4 μ m surface roughness.*
** *with 0.9 μm surface roughness.*

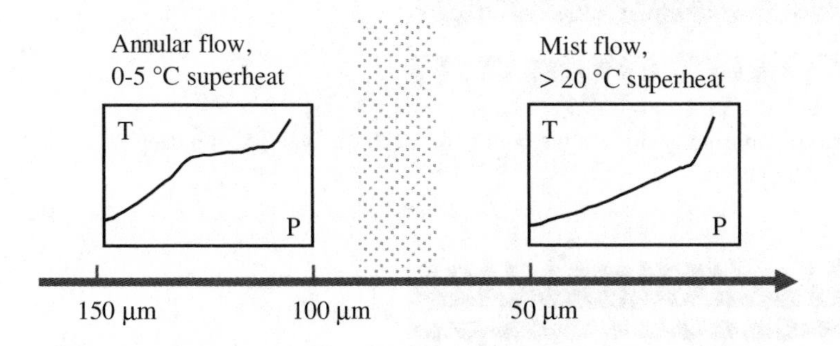

Fig. 6.12 Nucleate boiling and flow regime chart for 27-171 μm channels with plain walls. Boiling in channels larger than 100 μm diameter has low superheat and annular flow regime; boiling in channels smaller than 50 μm diameter has high superheat and mist flow regime; the channels between 50-100 μm shows transitional behavior.

6.4 Phase Change in Silicon Channels with Enhanced Walls

It has been noticed in the previous experiments that, the number of nucleation sites is very limited in the microchannels, even in the largest 171 μm diameter channel. This fact supports the new hypothesis that bubbles in these channels generate from irregular large-sized defects in the channel walls. In order to investigate the new hypothesis, channels with enhanced surfaces have been fabricated to conduct the same experiments as the plain-wall channels. The experimental results indicate that by using enhanced channel walls, the wall superheat at the onset of boiling is significantly lowered.

6.4.1 Nucleation on the Enhanced Surface

Experiments have been carried out with four channels with enhanced walls and hydraulic diameters ranging from 28-73 μm. Typical nucleate boiling was observed in all four channels with less than 5 °C wall superheat. Figure 6.13(a) is the image of bubble formation and growth from the bottom cavities in a 72.5 μm diameter channel. The bubble growth and departure cycles are exactly the same as what have been observed earlier in plain-wall channels larger than 100 μm; but the number of nucleation sites has significantly increased.

(a) Bubble nucleation from cavities in the bottom of a 120 μm wide, 52 μm deep, 72.5 μm hydraulic diameter channel.

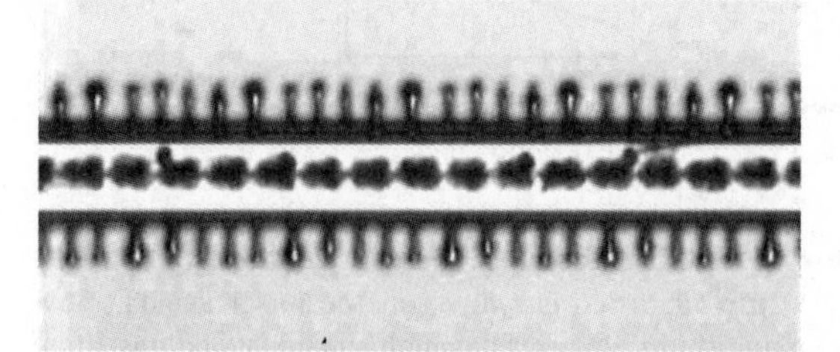

(b) Annular flow in a 74 μm wide, 52 μm deep, 61 μm hydraulic diameter channel at 0.1 ml/min DI water flow rate.

Fig. 6.13 Bubble nucleation and the annular flow pattern in enhanced-wall channels.

As the result of nucleate boiling with bubble growth and low superheat, annular flow pattern was induced after the onset of boiling, as shown in Fig. 6.13(b). The annular flow forms when there is enough number of nucleation sites and the bubble departure frequency is high enough. At the steady state, a continuous thin water film covers the entire channel wall, yielding much higher heat transfer coefficients as indicated in the wall temperature measurement in Fig. 6.15.

6.4.2 Wall Temperature Measurement During the Phase Change

The wall temperature change as a function of input heat rate at a constant liquid flow rate has been measured in the channels with enhanced walls. With 4-8 μm wide cavities in the channel, the wall superheat during the phase change is expected to be smaller than 5 °C from Hsu's model.

Figure 6.14 shows the wall temperature measurement results in a 47 μm diameter channel at 0.02 ml/min constant DI water flow rate. The channel dimensions are similar to the previously tested 44 μm channel with plain walls, and the DI water flow rate has been kept the same. Although steady annular flow was not observed in this channel, the wall superheat at the onset of boiling was successfully reduced to 1.5 °C, compared with the approximately 20 °C in Figs. 6.7, 6.9 and 6.11. The small temperature flat before the onset of boiling is the result of partial boiling.

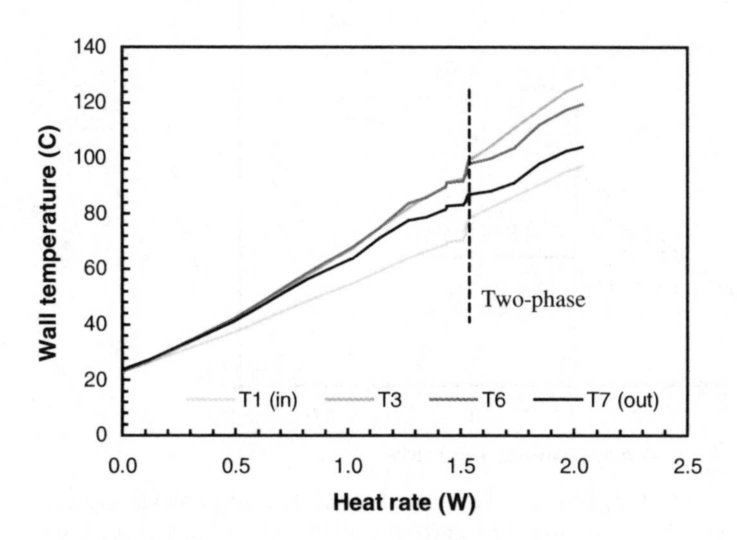

Fig. 6.14 Wall temperature change from four thermometers as a function of the heat rate in a 50 μm wide, 44 μm deep, 47 μm hydraulic diameter channel at 0.02 ml/min DI water flow rate. The thermometers have ±0.5 °C temperature accuracy.

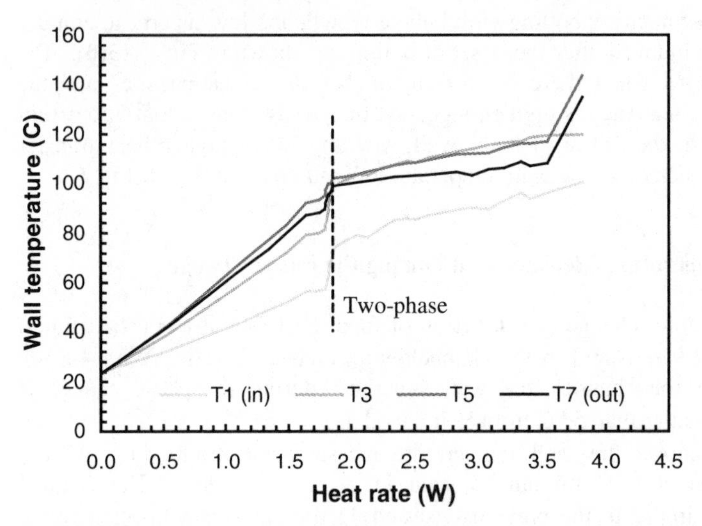

Fig. 6.15 Wall temperature change from four thermometers as a function of the heat rate in a 120 μm wide, 52 μm deep, 72.5 μm hydraulic diameter channel at 0.1 ml/min DI water flow rate. The thermometers have ±0.5 °C temperature accuracy.

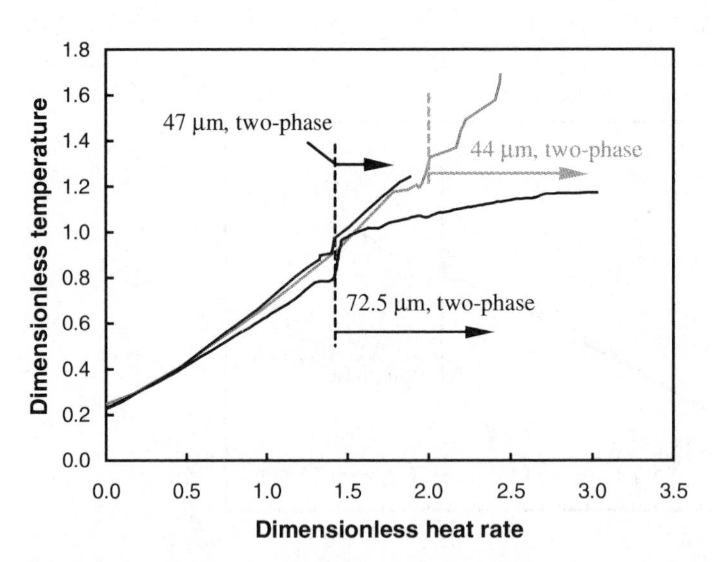

Fig. 6.16 Comparison of the boiling curves from plain-wall and enhanced-wall channels. One local wall temperature curve from a 44 μm plain-wall channel, a 47 μm enhanced-wall channel, and a 72.5 μm enhanced-wall channel are plotted in the dimensionless form. Because of enhanced boiling, the wall superheat has been successfully reduced in the 47 μm and 72.5 μm diameter channels.

Figure 6.15 shows the boiling curves from a 72.5 μm diameter channel at 0.1 ml/min constant DI water flow rate. The bubble nucleation has been observed in the cavities in the bottom of the channel and the onset of boiling occurs at 0.6 °C wall superheat. Because there were enough nucleation sites in the channel, annular flow developed in the steady state, yielding large heat transfer coefficients and a long temperature plateau in the two-phase flow region. After subtracting 40% heat loss in the experiment, the thermal resistance of 105 °C/W with the single-phase flow is reduced to 22 °C/W with the fully developed two-phase flow.

In the dimensionless form, the boiling curves at the highest local temperature spot from the 44 μm plain-wall channel in Fig. 6.7, the 47 μm enhanced-wall channel in Fig. 6.14, and the 72.5 μm enhanced-wall channel in Fig. 6.15 are compared in Fig. 6.16. With gas-trapping cavities, boiling in the two enhanced channels begins at almost the same point, indicating similar nucleate boiling conditions and mechanisms. Comparing with the boiling curve of the 44 μm channel, the wall enhancement brings the onset of boiling to much lower wall

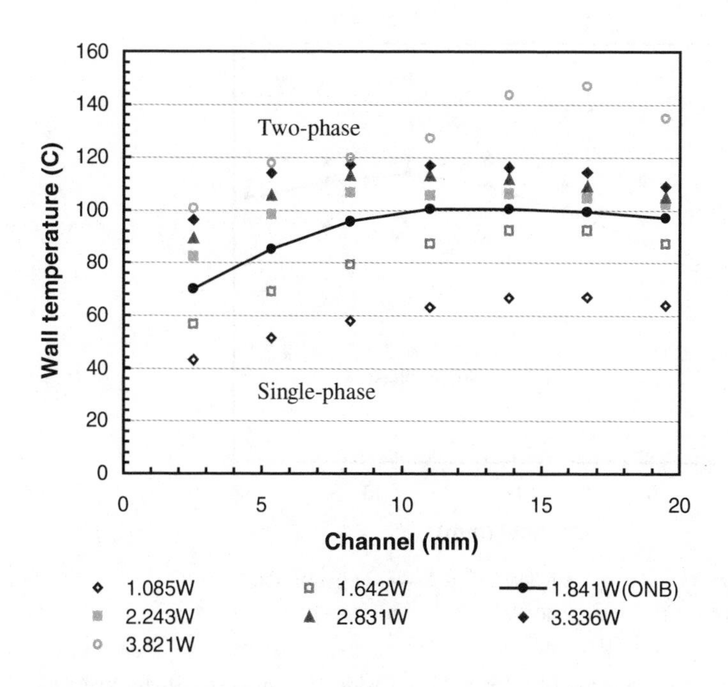

Fig. 6.17 Wall temperature profiles for a 120 μm wide, 52 μm deep, 72.5 μm hydraulic diameter channel at 0.1 ml/min DI water flow rate. The onset of boiling occurs at 1.841 W heat rate. The thermometers have ±0.5 °C temperature accuracy.

temperatures and normalized heat fluxes. Due to small dimensions and relatively large heat rate increment in the experiment, the transient annular flow quickly turned into mist flow in the 47 μm channel; hence the expected temperature plateau in the fully developed two-phase flow did not appear. However, the heat transfer can be improved in heat sinks with a large number of channels and better-controlled heat fluxes.

Figure 6.17 is the plot of wall temperature distributions at various input heat rates in the 72.5 μm channel at 0.1 ml/min constant DI water flow rate. The onset of boiling occurs at 1.841 W input heat rate, which is marked with a line. The highest wall temperature at the onset of boiling is 102.6 °C, only 0.2 °C higher than the boiling point of water at the measured channel pressure of 1.09 psi. The wall temperature distribution is expected to be linear in the single-phase region. The slight temperature reduction at the end of the channel is due to approximately 40% heat loss through conduction in the silicon chip. In the two-phase region, the wall temperature tends to become uniform, indicating that evaporation is the dominant form of heat transfer.

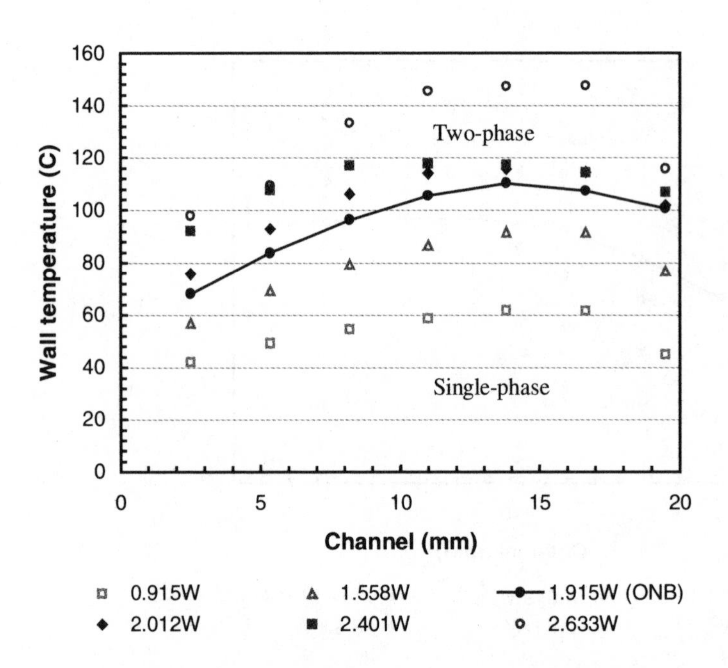

Fig. 6.18 Wall temperature profiles for a 74 μm wide, 52 μm deep, 61 μm hydraulic diameter channel at 0.1 ml/min DI water flow rate. The onset of boiling occurs at 1.915 W heat rate. The thermometers have ±0.5 °C temperature accuracy.

The same annular flow regime was observed in a 61 μm diameter channel. Figure 6.18 shows the wall temperature distributions with increasing input heat rate from this channel. The highest wall temperature at the onset of boiling is 110.4 °C, 1.6 °C higher than the boiling point of water at the measured channel pressure of 4.18 psi. Mist flow develops at 2.633 W heat rate, where a sharp temperature increase occurs due to the poor heat transfer coefficient associated with this flow regime.

6.4.3 Boiling Regime Chart for Enhanced-wall Channels

The wall temperature measurement results of the channels with enhanced walls are listed in Table 6.4. Although mist flow patterns can develop at moderate heat fluxes for channels smaller than 50 μm, the initiation of the phase change all occurs with less than 5 °C wall superheat, as predicted from heterogeneous nucleation theories for cavities ranging from 4-8 μm. Bubble growth has also been observed in all channels.

Table 6.4 Boiling regime summary in 28-73 μm microchannels with enhanced walls.

Hydraulic diameter (μm)	Channel width (μm)	Channel depth (μm)	Flow velocity (m/s)	Phase change mechanism/ flow regime	Wall super-heat (°C)
27.5	20	44	0.568	Nucleation/mist	0-5
47	50	44	0.152	Nucleation/mist	0-5
61	74	52	0.433	Nucleation/ annular	0-5
72.5	120	52	0.267	Nucleation/ annular	0-5

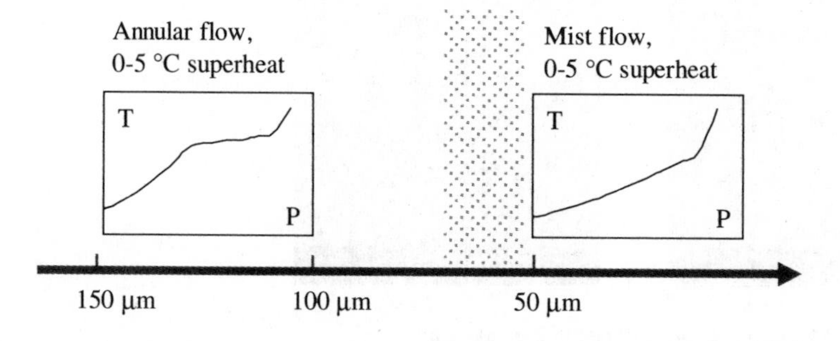

Fig. 6.19 Nucleate boiling and flow regime chart for microchannels with enhanced walls. With cavities in the channel walls, heterogeneous nucleation occurred even in the smallest

Figure 6.19 is the modified boiling mechanism and flow chart for sub-150 μm diameter microchannels with enhanced walls. Compared with Fig. 6.12, eruption boiling mechanism has been eliminated as the result of the existence of gas-trapping cavities in the channel walls. The chart supports the hypothesis that the eruption boiling and the unusual superheat associated with very small channels are due to the lack of nucleation sites. However, the limited internal space in very small channels determines that the mist flow will be easily induced due to the limited volume of the channel. As discussed earlier, one of the characteristics of the phase change in microchannels is that a single bubble can easily consume the entire channel space and induce either annular or mist flow depending on the vapor quality. Hence the quality must be carefully limited for microchannels below 50 μm diameters in heat sink design. Nevertheless, the nucleation mechanism can be controlled by making cavities in these channels, which have been proven effective in reducing the amount of wall superheat before the onset of boiling.

6.5 Discussion

The surface condition of the channel walls plays a key role in determining the nucleation mechanism in microchannels. The amount of wall superheat is closely related to the solid surface condition, but not very much affected by the liquid surface tension. The unusual wall superheat in smaller than 50 μm plain-wall channels can be eliminated with gas-trapping cavities, where Hsu's model is proven valid in microchannels for the first time. Some other interesting results and issues observed in the experiments are discussed below.

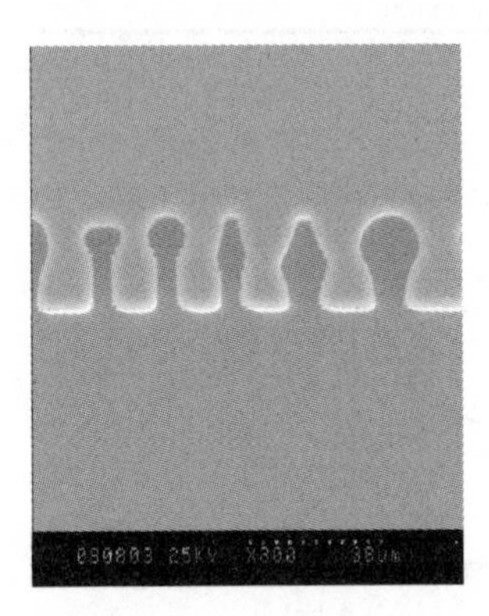

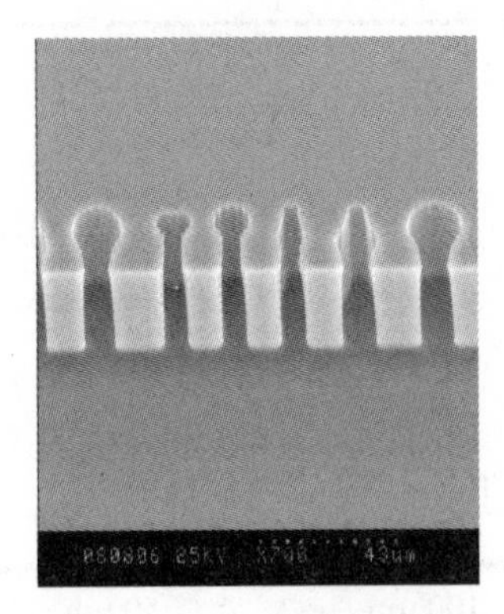

Fig. 6.20 SEM images of the notches in the side walls.

Figure 6.20 shows the SEM images of the notches in the side walls. The notches have 4×44 μm and 4×52 μm openings to the channel, and are 20 μm deep into the channel wall. They were originally designed as a comparison to the cavities in the bottom, which have dimensions on the same order of magnitude except smaller openings. Interestingly, although the bubbles prefer the side walls in a plain-wall channel, they "move" to the bottom cavities in the enhanced channels, and nucleation has never been observed in the side wall notches. Although surface features at this dimension are widely used as boiling enhancement structures in large boiling devices, they are not so effective in trapping gas when the notch dimension is on the same order of magnitude as the flow channel.

This fact supports Hsu's model of active nucleation sites in Equation (6.6), which defines a range of cavity sizes for a given thermal boundary layer thickness. The minimum cavity size model in Equation (6.5) is in agreement with Hsu's model for the lower bound, but it assumes an infinite thermal boundary layer, so there is no upper bound of cavity sizes. In Fig. 6.3, the notches in the microchannels fall in the region above the maximum cavity size for 10-50 μm thermal boundary layer thickness. In Hsu's model, these notches are considered cavities with infinite waiting period; hence they do not facilitate nucleate boiling.

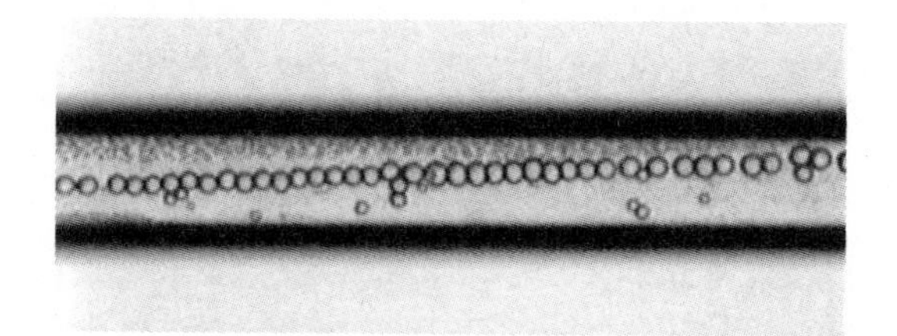

Fig. 6.21 Bubble formation in the 200 ppm TritonX-100 solution.

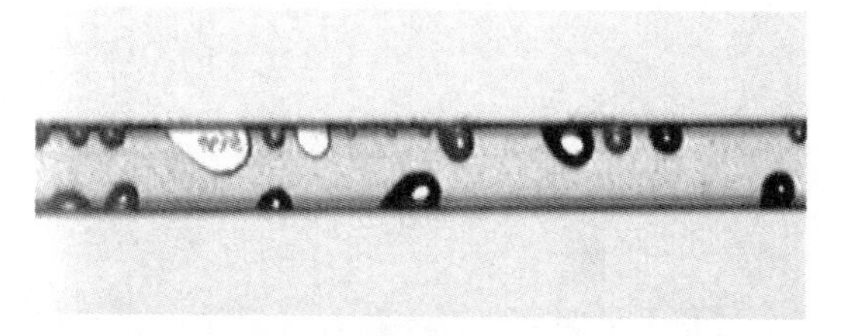

Fig. 6.22 Bubble nucleation in a 120 μm wide, 40 μm deep, 60 μm hydraulic diameter channel. Photoresist residues are found on the channel wall (not shown in the image).

Surface tension experiments prove that the unusual wall superheat is not directly caused by high liquid surface tensions. However, this does not necessarily suggest that surface tension is not related to the boiling mechanisms. Figure 6.21 shows an image of bubble formation in the channel. For unknown reasons, the bubbles form at low superheats, but they do not grow and do not initiate the phase change. Although not experimentally confirmed, the surface tension may be actually too small rather than too large. It is also possible that the surface tension has other effects that have not yet been noticed.

Although the number of nucleation sites is usually very limited in plain-wall channels, Fig. 6.22 shows an image of unusual nucleation in a 60 μm diameter channel without special surface modifications. However, some photoresist residues left from the fabrication process was found in the channel. The direct effect is that the density of nucleation sites is extremely high compared with other clean channels. A possible reason is that the residual photoresist makes the channel wall surface hydrophobic, while clean channel walls are hydrophilic due to a thin native oxide film on the silicon surface. In macroscale pool boiling, hydrophobic surfaces can facilitate bubble formation by providing more gas-trapping nucleation sites due to their poor wetting properties. This effect has been recently confirmed in microscale structures [6.15]. A postulate is that channels with hydrophobic surfaces will demonstrate different boiling regimes and enhance the nucleate boiling in small-dimension microchannels.

7 Conclusions

7.1 Phase Change in Sub-150 μm Diameter Microchannels

The phase change phenomena as well as two-phase heat transfer in microchannels with less than 150 μm hydraulic diameter have been discussed in this book. The discussions are based on various experiments that have been conducted to study the steady state and transient state characters of the phase change in microchannels. A unique experimental system with single freestanding microchannels and integrated heaters and thermometers has been introduced. These instrumented microchannels provide a reliable platform to carry out detailed, quantitative experiments on the boiling regimes in sub-150 μm diameter microchannels with the capability for simultaneous optical observation, thermal measurement, and transient pressure measurement. The overall behavior of the two-phase flow in silicon microchannels does not apparently depart from traditional theories in either heat transfer or boiling mechanisms, however, there are certain effects that are characteristic to microchannels associated with the scaling of the channel size. Specifically, because the internal area and volume of the microchannels are so small, negligible factors in large tubes become dominant, such as the condition of individual nucleation sites as well as the formation of individual bubbles.

7.1.1 Bubble Nucleation Mechanisms

Bubble nucleation during the phase change in microchannels takes the form of heterogeneous nucleation, and is governed by traditional theories. As in large tubes, vapor bubbles always form on the channel wall, usually from defective cavities, and continue to grow until they are carried away by the liquid. The unique phenomenon in microchannels is that large amounts of wall superheat can be easily induced due to the relatively smooth wall surface. And the unusual superheat directly causes sudden eruption of vapor at the onset of boiling, which is only observed in large tubes at extremely high heat fluxes.

Two nucleation mechanisms can be observed in sub-150 μm diameter silicon microchannels—the typical bubble nucleation with low superheat and vapor eruption with high degrees of superheat. Although the eruption boiling seems to be unique in microchannels, it is still a form of heterogeneous nucleation. Heterogeneous nucleation usually begins from gas-trapping cavities in a surface because they tend to provide the initial gas core to the vapor and become active nucleation sites. The size of the cavity plays a key role in the nucleation process, and the amount of wall superheat is a function of the cavity size. The micromachining techniques that are commonly used to fabricate silicon microchannels achieve an average surface roughness on the order of 0.1-1 μm, and very few defective cavities on the order of 2-5 μm. In sub-150 μm diameter microchannels with plain side walls, the active nucleation sites are usually the largest defective cavities in the channel wall. For the channels with larger than 100 μm diameters, there is normally a limited number of large-sized defective cavities, which becomes active nucleation sites at relatively low superheat, leading to the typical bubble nucleation process. On the other hand, when the channel diameter scales to less than 50 μm, the 2-5 μm diameter defective cavities are almost absent. As a result, the gas trapping nucleation sites in these channels are much smaller in size, sometime even well below 1 μm. These small-sized cavities require much higher degrees of superheat to become active nucleation sites with fast evaporation rates which cause vapor eruption. When 4-8 μm cavities are intentionally introduced into sub-50 μm diameter channels, the boiling mechanism changes back to normal bubble nucleation. In summary, the boiling mechanism is determined by the wall surface condition rather than liquid surface tension, and is still governed by traditional boiling theories.

Boiling enhancement can also be scaled into microchannels to some extent. Trenches and cavities on the order of 100 μm have been traditionally made into boiling devices to facilitate the nucleate boiling. These features cannot trap gas and serve as nucleation sites in microchannels any more because they are on the same order of magnitude as the channel itself, and actually become part of the channel. However, the wall superheat can be effectively controlled by creating less than 10 μm diameter cavities in the microchannels. The size range and the amount of wall superheat are in very good agreement with Hsu's model, which is for the first time used to explain bubble nucleation in microchannels.

7.1.2 Two-phase Flow Regimes in Microchannels

It is unique in microchannels that one bubble can easily occupy a large portion of the entire channel during the phase change. This phenomenon, although not a violation of traditional theories, causes quite a few differences in the two-phase regimes in microchannels. First, bubbly and slug flows that exist in large tubes are often absent in microchannels. Second, the two-phase flow regimes, primarily the annular flow and the mist flow, are more closely associated with the nucleation mechanism because the bubble generation process strongly affects the two-phase flow pattern. Third, in sub-50 μm diameter channels, mist flow seems to dominate.

Due to the limited internal space in sub-150 μm diameter channels, when the phase change begins, a single bubble quickly occupies the entire cross section of the channel and grows laterally towards the inlet and the outlet prior to departure.

In other words, there is no room for bubbly and slug flows to develop at the early phases of the phase change. Therefore, either annular flow or mist flow develops immediately after the onset of boiling. The heat flux and the local wall superheat determine which regime the two-phase flow appears in.

The annular flow can be only developed from bubble nucleation mechanism with less than 5 °C wall superheat and maintained at low to moderate vapor quality. The annular flow has the highest heat transfer coefficients in two-phase flow regimes in both macroscale and microscale channels. The condition of nucleation sites is critical to this flow regime in microchannels. Enhanced walls with gas-trapping cavities help induce the annular flow regime by facilitating the normal bubble nucleation process.

The mist flow can directly result from eruption nucleation mechanism with 20-50 °C wall superheat, due to relatively high heat fluxes required to activate the nucleation sites. It is also the final form of the two-phase flow when the heat flux or the vapor quality is high enough. However, due to the very limited volume of less than 50 μm diameter channels, a two-phase flow tends to shift to the mist flow regime even if an annular flow has already been established. This is because a single bubble growth can affect the flow condition in the entire channel, causing instabilities in the existing annular flow and quickly leading to the mist flow regime well before it is anticipated in large tubes.

7.1.3 Transient Characteristics

The small internal volume of sub-150 μm diameter microchannels also results in some transient phenomena, mainly because individual bubbles can affect some important parameters such as pressure and temperature. At the onset of the boiling, each bubble formation and departure causes transient pressure fluctuations on the order of 10-20 psi at 3-10 Hz frequencies. In addition, due to the required heating time at individual nucleation sites, whose dimensions dictate the amounts of wall superheat, up to 10 °C temporal local wall temperature increase may appear for a period of a few seconds.

7.1.4 Two-phase Heat Transfer Model for Sub-100 μm Channels

A two-phase flow and heat transfer model, which predicts the wall temperature distribution and the pressure drop at any given heat flux and liquid flow rate, has been introduced to facilitate the design of microchannel heat sinks. The model consists of a homogeneous flow model and energy equations with Kandlikar's correlation, and is in good agreement with the heat transfer measurements from 28-60 μm plasma etched silicon channels with plain side walls. Kandlikar's correlation, one of the most widely accepted traditional heat transfer coefficient correlations, is first proven valid in channels below 500 μm hydraulic diameter. The model has been used to optimize the design of a heat sink that removes 200 W from a 1×1 inch IC chip within 80 °C maximum temperature rise, which will be introduced in Sect. 7.2.

7.1.5 General Design Rules for Two-phase Microchannel Heat Sinks

Although the two-phase flow behavior is still governed by traditional boiling theories, some characteristic phenomena exist in sub-150 μm diameter channels due to the small channel dimensions. Some special issues must be considered in the heat sink design.

Despite that a high vapor quality means that more heat is absorbed by the liquid-vapor phase change, it may not benefit the microchannel heat sinks. In the two-phase forced convective boiling, the heat transfer coefficient or the thermal resistance is the major figure of merit. This requires that the heat sink operate primarily under the annular flow regime due to its highest heat transfer coefficients. However, mist flow can easily develop at qualities that are still considered low in large tubes because of more instability factors in microchannels. Therefore, the two-phase flow quality and how the flow patterns transform must be carefully considered. The recommended target quality is 0.1-0.3 for 50-150 μm diameter channels. This number is based on the simulation results that are in good agreement with the measured channel wall temperatures.

Pressure drop along the channels is another major consideration. In the single phase forced convection, because the heat transfer coefficient is independent of the flow velocity, larger pressure drop yields larger amount of heat transfer. However, in the two-phase heat transfer, the wall must be superheated to a certain degree above the boiling point of the liquid. That is, the boiling point under a certain pressure must be taken into consideration—if the boiling point increases linearly or even exponentially with the ambient pressure, the pressure drop along the channel must be limited in the design. As an example, if water is used as the working fluid, the pressure drop should be controlled below 2.5 psi to avoid more than 5 °C rise in the boiling point.

In addition to the global pressure drop along the channels, local transient pressure pulses due to bubble nucleation in each channel are on the order of 10-20 psi, even larger than the designed pressure drop along the channel. Although the fast transient fluctuations have not been found to affect the two-phase flow or heat transfer, they may be destructive and cause device failure. Based on the experimental results, the spacing between channels should be greater than 100 μm to provide enough support area.

The optimum channel diameter is larger than 100 μm for a steady annular two-phase flow regime. The boiling regime charts in Figs. 6.11 and 6.18 are based on experimental results with 27-171 μm diameter microchannels, on how the surface conditions affect the two-phase flow regimes. Wall superheat and flow regimes are the most important properties for the phase change in microchannels. For channels smaller than 100 μm hydraulic diameter, either enhanced surfaces such as cavities or other surface treatment must be used in order to prevent large amounts of wall superheat. Even in larger channels, enhanced wall surfaces help to achieve steady nucleate boiling with annular flows. However, microchannels with smaller than 20 μm diameter may not be suitable for heat sinks because more than 10 °C wall superheat is expected in these channels. Hsu's model for active nucleation sites indicates that a smaller diameter channel itself causes higher degrees of superheat, because of its thinner thermal boundary layer. For example, if a 5 μm diameter channel is used, the minimum wall superheat will be 25 °C, with 0.4-0.8 μm cavities as potential nucleation sites. Also in very small diameter channels, large heat fluxes are required to activate their small-sized nucleation sites. The heat fluxes are usually high enough to quickly turn the two-phase flow into mist flow regime, yielding poor heat transfer rates. As a conclusion, even for "microchannel heat sinks", the channels should take the maximum possible sizes.

7.2 A Sample Design of a Two-phase Microchannel Heat Sink

On the basis of the two-phase flow and heat transfer model introduced in Chap. 4, a microchannel heat sink which is capable of removing 200 W from a 1×1 inch IC chip has been designed and optimized [4.14]. Given 15 ml/min water flow rate, which is expected from an electroosmotic pump, the design parameters including the channel dimensions and the number of the channels are optimized with simulations. The figure of merit is the steady state wall temperature under two-phase cooling.

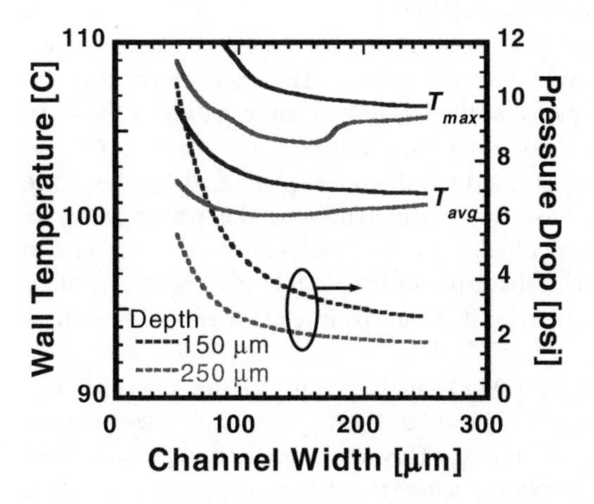

Fig. 7.1 Optimization of a 200 W microchannel heat sink. The figure of merit of the design is the wall temperature under the two-phase cooling. By courtesy of Jae-Mo Koo and Linan Jiang.

Considering the thickness of a silicon wafer (500 μm) and the possible large pressure increase in the two-phase heat transfer, the maximum channel depth is limited to 250 μm. To ensure the bonding strength between the silicon substrate and the glass slide, the channel wall thickness, or the spacing between two channels, is fixed at 100 μm. Wall temperature and pressure drop for channels with widths ranging from 50 to 250 μm are calculated and the results are shown in Fig. 7.1. The solid lines in the figure are the maximum and average chip temperatures for varying channel dimensions. The results for 150 μm and 250 μm deep channels with maximum possible number of channels are plotted as comparisons. The lowest wall temperature is obtained from a 100-channel design, with individual channels of 250 μm deep and 150 μm wide. The dashed lines are the pressure drop curves for 150 μm and 250 μm channel depth respectively. The curves show that the expected pressure drop for the lowest temperature point is 2.8 psi, which raises the boiling point of water to 105.9 °C. The heat sink is expected to remove 200 W with 100 °C average chip temperature and 106 °C maximum local wall temperature at 15 ml/min water flow rate supplied with an electroosmotic pump.

7.3 Future Studies

Phase change phenomena in sub-150 μm hydraulic diameter channels have been discussed in this book. Questions such as how the boiling regimes and forced convective heat transfer in microchannels differ from large boiling tubes have been answered. However, there are still some unknowns in this field. Some directions for future studies are proposed as follows:

Both homogeneous and annular flow models have been used in Chap. 4 when comparing the modeling and experimental results. The data show that the homogeneous model, which assumes that the liquid and vapor moves at the same velocity, is better supported by the experiments. However, the channels used in the heat transfer measurements were 28-60 μm diameter, plain-wall devices. The later visualized flow pattern is mist flow, which may be the reason why the homogeneous model is better supported. More simulations are necessary to examine the modeling accuracy with the experimental results from known annular flows as observed in larger than 100 μm diameter plain-wall channels as well as smaller enhanced-wall channels.

The boiling mechanism in plain-wall channels below 50 μm diameter was determined to be heterogeneous nucleation from the wall temperature measurements without visual confirmations. This could be very important if the nucleation process, the vapor eruption, actually behaves differently in these channels. More efforts with high-speed photography and high-frequency thermometry are necessary for further studies.

As the initial experiments, the current dimensions of the nucleation cavities are not optimized, because the percentage of observed active nucleation sites over the total number of cavities is less than 20%. Since the enhanced-wall channel experiments have been in very good agreement with Hsu's model, simulations from this model can be used to find the best range of cavity size in heat sink designs. On the other hand, the surface wetting properties may also affect the nucleation sites in microchannels. Hydrophobic surfaces have been reported to facilitate the bubble nucleation process. In microchannels, hydrophobic surfaces tend to trap air when the liquid enters the channel, hence provide more active nucleation sites. The study of channels with varying surface wettability may lead to new methods of boiling enhancement in microchannels.

Finally, Hsu's model suggests that at least 10 °C wall superheat is required for nucleate boiling in smaller than 20 μm diameter channels. This prediction has not been confirmed by experimental results. Although the experiments in this book agree well with Hsu's model for 28-73 μm diameter enhanced-wall channels, interesting phenomena still could happen in even smaller dimensions. Because Hsu's model considers the boundary layer development in pool boiling, whose temperature profile and correlations are expected to be different from those of internal flows, especially in microchannels, it is possible that there exists a transition in smaller than 20 μm diameter channels where the phase change shows a departure from classical theories. In fact, along with the scaling down of channel dimensions, the surface tension is also expected to become a dominant force at some threshold, but this effect has not been observed in 27-171 μm diameter channels. Therefore, less than 20 μm diameter channels should draw attention in the future studies.

Appendix A: Process Flow Chart for Single-channel Devices with Combined Heater and Thermometers

Step	Process	Parameters	Notes
1	Lithography	1 µm Shipley 3612 positive photoresist	Alignment marks
2	Silicon etch	5000 Å	
3	Photoresist strip		
4	Oxidation	Wet 850 °C, 13min, (~250 Å)	Implantation protection
5	Lithography	1 µm Shipley 3612	Resistors
6	Ion implantation	Boron, 1E16 at 40 keV	
7	Photoresist strip		
8	HF etch	20:1 BOE, 1min	
9	Oxidation	Wet 1100 °C, 15min (~2500 Å)	Resistor passivation
10	Annealing	1150 °C, 6 hours (~7 µm junction depth)	PN junction deepening
11	Lithography	1 µm Shipley 3612	
12	HF etch	6:1 BOE, 3min	Contact windows
13	Photoresist strip		
14	Al deposition	99% Al, 1% Si, 1 µm	
15	Lithography	1.6 µm Shipley 3612	Al contacts
16	Al etch	Al etch, 4min	
17	Freckle etch	20s	
18	Photoresist strip		
19	FGA	400 °C, 45min	
20	LTO	400 °C, 2000 Å	
21	Lithography	1.6 µm Shipley 3612	Contact pads
22	Oxide etch	Dry etch	
23	Photoresist strip		
24	Lithography	7 µm SPR220-7 positive photoresist	Reservoir
25	Silicon etch	DRIE silicon etch	
26	Photoresist strip		
27	Lithography	7 µm SPR220-7	Channel

28	Silicon etch	DRIE	
29	Photoresist strip		
30	Wafer bonding	1.6 µm photoresist, 1 hour bake at 110 °C	Support wafer for through etch
31	Lithography	10 µm SPR220-7	Inlet/outlet holes
32	Silicon etch	DRIE	
33	Photoresist strip		
34	Wafer unbonding	Immersion in acetone, ~3 hours	Remove the support wafer
35	Anodic bonding	Silicon to glass slide, 1200 V, 250 °C	

Appendix B: Process Flow Chart for Single-channel Devices with Separate Heater and Thermometers

Step	Process	Parameters	Notes
1	Lithography	1 µm Shipley 3612 positive photoresist	Alignment marks
2	Silicon etch	5000 Å	
3	Photoresist strip		
4	Oxidation	Wet 850 °C, 13min, (~250 µm)	Implantation protection
5	Lithography	1 µm Shipley 3612	Resistors
6	Ion implantation	Boron, 1E15 at 40 keV	
7	Photoresist strip		
8	HF etch	20:1 BOE, 1min	
9	Oxidation	Wet 950 °C, 1 hour (~2500 Å)	Resistor annealing and passivation
10	Lithography	1 µm Shipley 3612	
11	HF etch	6:1 BOE, 3min	Contact windows
12	Photoresist strip		
13	Al deposition	99% Al, 1% Si, with pre-sputtering etch, 2 µm	
14	Lithography	1.65 µm Shipley 3612	Al track and contacts
15	Al etch	Al etch, 8min	
16	Freckle etch	20s	
17	Photoresist strip		
18	FGA	400 °C, 45min	
19	LTO	400 °C, 2000 Å	
20	Lithography	1.65 µm Shipley 3612	Contact pads
21	Oxide etch	Dry etch	
22	Photoresist strip		
23	Lithography	7 µm SPR220-7 positive photoresist	Reservoir
24	Silicon etch	DRIE silicon etch	
25	Photoresist strip		

26	Lithography	7 µm SPR220-7	Channel
27	Silicon etch	DRIE	
28	Photoresist strip		
29	Wafer bonding	1.6 µm photoresist, 1 hour bake at 110 °C	Support wafer for through etch
30	Lithography	7 µm SPR220-7	Inlet/outlet holes
31	Silicon etch	DRIE	
32	Photoresist strip		
33	Wafer unbonding	Immersion in acetone, ~3 hours	Remove the support wafer
34	Anodic bonding	Silicon to glass slide, 1200 V, 250 °C	

References

[1.1] Moore G. E., "Cramming More Components onto Integrated Circuits", *Proceedings of the IEEE*, Jan. 1998, Vol. 86, No. 1, pp. 2-85, reprinted from *Electronics*, Apr. 1965, Vol. 38, No. 8, pp. 114-117.

[1.2] Intel product information, http://www.intel.com/.

[1.3] Patel C. D., "Enabling Pumped Liquid Loop Cooling: Justification and the Key Technology and Cost Barriers", *Proceedings of 2000 International Conference on High-Density Interconnect and Systems Packaging*, Apr. 2000, pp. 145-152.

[1.4] International Technology Roadmap for Semiconductors, http://public.itrs.net/.

[1.5] Morgan M. J., Chang W. S., Pais M. R., Chow L. C., "Comparison of High Heat-Flux Cooling Applications", *Proceedings of SPIE High Heat Flux Engineering*, 1992, Vol. 1739, pp. 17-28.

[1.6] Zhou P., Touzelbaev M., Jiang L., Hayward J., Goodson K. E., "Compact Closed-Loop Forced Convective Cooling Systems for Integrated Circuits", submitted to *IEEE Transactions on Reliability*.

[1.7] Mudawar I., "Direct-Immersion Cooling for High Power Electronic Chips", *Proceedings of Inter Society Conference on Thermal Phenomena*, Feb. 1992, pp. 74-84.

[1.8] Mudawar I., "Assessment of High-Heat-Flux Thermal Management Schemes", *IEEE Transactions on Components and Packaging Technologies*, Jun. 2001, Vol. 24, No. 2, pp. 122-141.

[1.9] Arik M., Bar-Cohen, A., "Immersion Cooling of High Heat Flux Microelectronics with Dielectric Liquids", *Proceedings of 4th International Symposium on Advanced Packaging Materials Processes, Properties and Interfaces*, Mar. 1998, pp. 229-247.

[1.10] Bhavnani S. H., Fournelle G., Jaeger R. C., "Immersion-Cooled Heat Sinks for Electronics: Insight from High-Speed Photography", *IEEE Transactions on Components and Packaging Technologies*, Jun. 2001, Vol. 24, No. 2, pp. 166-176.

[1.11] Tian S-R., Takamatsu H., Honda H., "Experimental Study on the Immersion Cooling of an Upward-Facing Multichip Module with an Opposing Condensing Surface", *Heat Transfer-Janpanese Research*, 1998, Vol. 27, No. 7, pp. 497-508.

[1.12] Meyer G., "Heat Pipe Cooling for Notebook Computers", *Proceedings of 3rd Annual Portable by Design Conference*, Mar. 1996, pp. 549-554.

[1.13] Mochizuki M., Saito Y., Goto K., Nguyen T., Ho P., Malcolm M, Morando M. P., "Hinged Heat Pipes for Cooling Notebook PCs", *Proceedings of 13th Annual IEEE Semiconductor Thermal Measurement and Management Symposium*, Jan. 1997, pp. 64-72.

[1.14] Nguyen T., Mochizuki M., Mashiko K., Saito Y., Sauciuc L., Boggs R., "Advanced Cooling System Using Miniature Heat Pipes in Mobile PC", *Proceedings of 6th Intersociety Conference on Thermal and Thermomechanical Phenomena in Electronic Systems*, May 1998, pp. 507-511.

[1.15] Ali A., DeHoff R., Grubb K., "Advanced Heat Pipe Thermal Solutions for Higher Power Notebook Computers", *Proceedings of 5th Annual Pan Pacific Microelectronics Symposium*, Jan. 2000, pp. 143-148.

[1.16] Zuo Z. J., North M. T., Wert K. L., "High Heat Flux Heat Pipe Mechanism for Cooling of Electronics", *IEEE Transactions on Components and Packaging Technologies*, Jun. 2001, Vol. 24, No. 2, pp. 220-225.

[1.17] Kirhberg J., Yerkes K., Trebotich D., Liepmann D., "Cooling Effect of a MEMS Based Micro Capillary Pumped Loop for Chip-Level Temperature Control", *Proceedings of ASME IMECE, MEMS-Vol. 2,* Nov. 2000, pp. 143-150.

[1.18] Mukherjee S., Mudawar I., "Smart, Low-Cost, Pumpless Loop for Micro-Channel Electronic Cooling Using Flat and Enhanced Surfaces", *Proceedings of the ASME 8th Intersociety Conference on Thermal and Thermomechanical Phenomena in Electronic Systems*, May 2002, pp. 360-370.

[1.19] Tuckerman D. B., Pease R. F. W., "High-Performance Heat Sinking for VLSI", *IEEE Electron Device Letters*, May 1981, Vol. EDL-2, No. 5, pp. 126-129.

[1.20] Kiper A. M., "Impinging Water Jet Cooling of VLSI Circuits", *International Communications in Heat and Mass Transfer*, Nov. 1984, Vol. 11, pp. 517-526.

[1.21] Ma C. F., Bergles A. E., "Jet Impingement Nucleate Boiling", *ASME Heat Transfer in Electronic Equipment*, 1983, Vol. 28, pp. 5-12.

[1.22] Estes K. A., Mudawar I., "Comparison of Two-Phase Electronic Cooling Using Free Jets and Sprays", *Proceedings of ASME Advances in Electronic Packaging*, Mar. 1995, EEP-Vol. 12, pp. 975-987.

[1.23] Ortiz L., Gonzalez J. E., "Experiments on Steady State High Heat Fluxes Using Spray Cooling", *Experimental Heat Transfer*, Jul. 1999, Vol. 12, No. 3, pp. 215-233.

[1.24] Yang J., Pais M. R., Chow L. C., "High Heat Flux Spray Cooling", *SPIE Vol. 1739, High Heat Flux Engineering*, 1992, pp. 29-40.

[1.25] Wadsworth D. C., Mudawar I., "Cooling of a Multichip Electronic Module by Means of Confined Two-Dimensional Jets of Dielectric Liquid", *Journal of Heat Transfer*, Nov. 1999, Vol. 112, pp. 891-898.

[1.26] Lagorce L., Kercher D., English J., Brand O., Glezer A., Allen M., "Batch-Fabricated Microjet Coolers for Electronic Components", *Proceedings of 1997 Internatinal Symposium on Microelectronics*, Oct. 1997, pp. 494-499.

[1.27] Stefanescu S., Mehregany M., Leland J., Yerkes K., "Micro Jet Array Heat Sink for Power Electronics", *IEEE MEMS Workshop*, Jan. 1999, pp. 165-170.

[1.28] Lin Q., Wu S., Yuen Y., Tai Y-C., Ho C-M., "MEMS Impinging-Jet Cooling", *Proceedings of ASME IMECE*, Nov. 2000, pp. 137-142.

[1.29] Zhang L., Wang E. N., Koo J-M., Jiang L., Goodson K. E., Santiago J. G., Kenny T. W., "Microscale Liquid Jet Impingement", *Proceedings of ASME IMECE*, Nov. 2001, Vol. 2, MEMS-23820.

[1.30] Wang E. N., Zhang L., Koo J-M., Jiang L., Goodson K. E., Santiago J. G., Kenny T. W., "Microscale Liquid Jet Impingement", *Technical Digest, Solid State Sensors and Actuators Workshop*, Jun. 2002, pp. 46-49.

[1.31] Mudawar I., Maddox D. E., "Enhancement of Critical Heat Flux from High Power Microelectronic Heat Sources in a Flow Channel", *Transactions of ASME, Journal of Electronic Packaging*, Sep. 1990, Vol. 112, No. 3, pp. 241-248.

[1.32] Zeng S., Chen C.-H., Mikkelsen J. C. Jr., Santiago J. G., "Fabrication and Characterization of Electrokinetic Micro Pumps", *Proceedings of the ASME ITHERM*, May 2000,Vol. 2, pp. 31-36.

[1.33] Chen C.-H., Mikkelsen J. C. Jr., Santiago J. G., "Development of a Planar Electrokinetic Micropump", *Proceedings of the ASME IMECE*, Nov. 2000, Vol. 2, pp. 523-528.

[1.34] Laser D. J., Yao S., Chen C.-H., Mikkelsen J. C. Jr., Goodson K. E., Santiago J. G., Kenny T. W., "A Micromachined Silicon Low-Voltage Parallel-Plate Electrokinetic Pump", *Proceedings of Transducers'01*, Jun. 2001, pp. 920-923.

[1.35] Jiang L., Mikkelsen J., Koo J-M., Huber D., Yao S., Zhang L., Zhou P., Maveety J. G., Prasher R., Santiago J. G., Kenny T. W., Goodson K. E., "Closed-Loop Electroosmotic Microchannel Cooling System for VLSI Circuits", *IEEE Transaction on Components and Packaging Technologies*, Sep. 2002, Vol. 25, No. 3, pp. 347-355.

[1.36] Phillips R. J., "Microchannel Heat Sinks", *The Lincoln Laboratory Journal*, 1988, Vol. 1, No. 1, pp. 31-48.

[1.37] Gillot C., Schaeffer C., Bricard A., "Integrated Micro Heat Sink for Power Multichip Module", *IEEE Transactions on Industry Applications*, Jan. 2000, Vol. 36, No. 1, pp. 217-221.

[1.38] Hsu T. R., Bar-Cohen A., Nakayama W., "Manifold Micochannel Heat Sinks: Theory and Experiment", *Proceedings of the ASME International Electronic Packaging Conference*, Mar. 1995, Vol. 2, pp. 829-835.

[1.39] Hahn R., Kamp A., Ginolas A., Schmidt M., Wolf J., Glaw V., Topper M., Ehrmann O., Reichl H., "High Power Multichip Modules Employing the Planar Embedding Technique and Microchannel Water Heat Sinks", *Proceedings of IEEE 13th SEMI-THERM Symposium*, May 1997, pp. 49-56.

[1.40] Martin P. M., Bennett W. D., Johnston J. W., "Microchannel Heat Exchangers for Advanced Climate Control", *Proceedings of SPIE*, 1995, Vol. 2639, pp. 82-88.

[1.41] Vydai A. V., Soshelev S. B., Reznikov G. V., Kharitonov V. V., Cheremushkin S.V., "Thermophysical Design of the Parameters of a Computer Board with a Microchannel Cooling System", *Journal of Engineering Physics and Thermophysics*, Jan. 1993, Vol. 64, No. 1, pp. 80-86.

[1.42] Missaggia L. J., Walpole J. N., Liau Z. L. Phillips R. J., "Microchannel Heat Sinks for Two-Dimensional High-Power-Density Diode Laser Arrays", *IEEE Journal of Quantum Electronics*, Sep. 1989, Vol. 25, No. 9, pp. 1988-1992.

[1.43] Mundinger D., Beach R., Benett W., Solarz R., Krupke W., Staver R., Tuckerman D., "Demonstration of High-Performance Silicon Microchannel Heat Exchangers for Laser Diode Array Cooling", *Applied Physics Letter*, Sep. 1988, Vol. 53, No. 12, pp. 1030-1032.

[1.44] Beach R., Benett W. J., Freitas B. L., Mundinger D., Comaskey B. J., Solarz R. W., Emanuel M. A., "Modular Microchannel Cooled Heatsinks for High Average Power Laser Diode Arrays", *IEEE Journal of Quantum Electronics*, Apr. 1992, Vol. 28, No. 4, pp. 966-976.

[1.45] Koo J-M., Jiang L., Bari A., Zhang L., Wang E., Kenny T. W., Santiago J. G., Goodson K. E., "Convective Boiling in Microchannel Heat Sinks with Spatially-Varying Heat Generation", *Proceedings of the ASME 8th Intersociety Conference on Thermal and Thermomechanical Phenomena in Electronic Systems*, May 2002, pp. 341-346.

[2.1] White F. M, *Fluid Mechanics*, 4th Edition, McGraw-Hill Companies, Inc., 1999, pp. 338-344.

[2.2] Incropera F. P., DeWitt D. P., *Heat and Mass Transfer*, 4th Edition, John Wiley & Sons Inc., 1996, pp. 420-461.

[2.3] Bowers M. B., Mudawar I., "High Flux Boiling in Low Flow Rate, Low Pressure Drop Mini-Channel and Micro-Channel Heat Sinks", *International Journal of Heat and Mass Transfer*, 1994, Vol. 37, No. 2, pp. 321-332.

[2.4] Urbanek W., Zemel J., Bau H. H., "An Investigation of the Temperature Dependence of Poiseuille Numbers in Microchannel Flow", *Journal of Micromechanics and Microengineering*, Dec. 1993, Vol. 3, No. 4, pp. 206-209.

[2.5] Zeighami R., Laser D., "Experimental Investigation of Flow Transition in Microchannels Using Micron-Resolution Particle Image Velocimetry", *Proceedings of the ASME ITHERM*, May 2000,Vol. 2, pp. 148-153.

[2.6] Tso C. P., Mahulikar S. P., "The Role of the Brinkman Number in Analyzing Flow Transitions in Microchannels", *International Journal of Heat and Mass Transfer*, 1999, Vol. 42, pp. 1813-1833.

[2.7] Jacobi A. M., "Flow and Heat Transfer in Microchannels Using a Microcontinuum Approach", *Transaction of the ASME Journal of Heat Transfer*, Nov. 1989, Vol. 111, No. 4, pp. 1083-1085.

[2.8] Rahman M. M., Gui F., "Experimental Measurements of Fluid Flow and Heat Transfer in Microchannel Cooling Passages in a Chip Substrate", *Proceedings of the ASME Electronic Packaging Conference*, Sep. 1993, Vol. 2, pp. 685-692.

[2.9] Peng X. F., Wang H. B., Peterson G. P., Ma H. B., "Experimental Investigation of Heat Transfer in Flat Plates with Rectangular Microchannels", *International Journal of Heat and Mass Transfer*, 1995, Vol. 38, No. 1, pp. 127-137.

[2.10] Jiang L., Wong M., Zohar Y., "Phase Change in Microchannel Heat Sinks with Integrated Temperature Sensors", *Journal of MEMS*, Dec. 1999, Vol. 8, No. 4, pp. 358-365.

[2.11] Lienau J. J., "The Recommended Geometry for Laminar Microchannel Heat Sinks", *Proceedings of the ASME CAE/CAD and Thermal Management Issues in Electronic Systems*, Nov. 1997, EEP-Vol. 23/HTD-Vol. 356, pp. 47-54.

[2.12] Samalam V. K., "Convective Heat Transfer in Microchannels", *Journal of Electronic Materials*, Sep. 1989, Vol. 18, No. 5, pp. 611-617.

[2.13] Knight R. W., Hall D. J., Goodling J. S., Jaeger R. C., "Heat Sink Optimization with Application to Microchannels", *IEEE Transactions of Components, Hybrids, and Manufacturing Technology*, Oct. 1992, Vol. 15, No. 5, pp. 832-842.

[2.14] Qu W., Mudawar I., "Analysis of Three-Dimensional Heat Transfer in Micro-Channel Heat Sinks", *International Journal of Heat and Mass Transfer*, Sep. 2002, Vol. 45, No. 19, pp. 3973-3985.

[2.15] Hetsroni G, Mosyak A., Segal Z., Ziskind G., "A Uniform Temperature Heat Sink for Cooling of Electronic Devices", *International Journal of Heat and Mass Transfer*, Jul. 2002, Vol. 45, No. 16, pp. 3275-3286.

[2.16] Zhao C. Y., Lu T. J., "Analysis of Microchannel Heat Sinks for Electronics Cooling", *International Journal of Heat and Mass Transfer*, Nov. 2002, Vol. 45, No. 24, pp. 4857-4869.

[2.17] Lee P. S., Ho J. C., Xue H., "Experimental Study on Laminar Hear Transfer in Microchannel Heat Sink", *Proceedings of the ASME 8th Intersociety Conference on Thermal and Thermomechanical Phenomena in Electronic Systems*, May 2002, pp. 379-386.

[2.18] Judy J., Maynes D., Webb B. W., "Characterization of Frictional Pressure Drop for Liquid Flows Through Microchannels", *International Journal of Heat and Mass Transfer*, Aug. 2002, Vol. 45, No. 17, pp. 3477-3489.

[4.1] Poh S. T., Ng E. Y. K., "Heat Transfer and Flow Issues in Manifold Microchannel Heat Sinks: a CFD Approach", *Proceedings of the 2nd Electronics Packaging Technology Conference*, Dec. 1998, pp. 246-250.

[4.2] Kim S. J., Kim D., Lee D. Y., "On the Local Thermal Equilibrium in Microchannel Heat Sinks", *International Journal of Heat and Mass Transfer*, 2000, Vol. 43, pp. 1735-1748.

[4.3] Ambatipudi K. K., Rahman M. M., "Analysis of Conjugate Heat Transfer in Microchannel Heat Sinks", *Numerical Heat Transfer*, 2000, Part A, No. 37, pp. 711-731.

[4.4] Carey V., *Liquid-Vapor Phase-Change Phenomena*, Hemisphere Publishing Corporation, 1992, pp. 520-524.

[4.5] Lee H. J., Lee S. Y., "Pressure Drop Correlations for Two-Phase Flow within Horizontal Rectangular Channels with Small Heights", *International Journal of Multiphase Flow*, 2001, Vol. 27, pp. 783-796.

[4.6] Stanley R. S., Barron R. F., "Two-Phase Flow in Microchannels", *Proceedings of the ASME IMECE*, Nov. 1997, DSC-Vol.62/HTD-Vol. 354, pp. 143-152.

[4.7] Riehl R. R., Selehim P. Jr., Ochterbeck J. M., "Comparison of Heat Transfer Correlations for Single- and Two-Phase Microchannel Flows for Microelectronics Cooling", *Proceedings of the ASME 6th Intersociety Conference on Thermal and Thermomechanical Phenomena in Electronic Systems*, May 1998, pp. 409-416.

[4.8] Peles Y. P., Yarin L. P., Hetsroni G., "Steady and Unsteady Flow in a Heated Capillary", *International Journal of Multiphase Flow*, 2001, Vol. 27, pp. 577-598.

[4.9] Jacobi A. M., Thome J. R., "Heat Transfer Model for Evaporation of Elongated Bubble Flows in Microchannels", *Transactions of the ASME Journal of Heat Transfer*, Dec. 2002, Vol. 124, No. 6, pp. 1131-1136.

[4.10] Ryu J. H., Choi D. H. Kim S. J., "Numerical Optimization of the Thermal Performance of a Microchannel Heat Sink", *International Journal of Heat and Mass Transfer*, Jun. 2002, Vol. 45, No. 13, pp. 2823-2827.

[4.11] Chen N. C. J., Fekle D. K., Yoder G. L., "Thermal Analysis of Two-Phase Microchannel Cooling", *Proceedings of ASME IMECE*, Nov. 1996, pp. 137-143.

[4.12] Hoffman M. A., Stetson J. D., "Modeling of Convective Subcooled Boiling in Microtubes for High Heat Fluxes", *Proceedings of SPIE High Heat Flux Engineering*, 1992, Vol. 1739, pp. 60-77.

[4.13] Incropera F. P., DeWitt D. P., *Heat and Mass Transfer*, 4th Edition, John Wiley & Sons Inc, 1996, pp. 829.

[4.14] Koo J-M., Jiang L., Zhang L., Zhou P., Banerjee S. S., Kenny T. W., Santiago J. G., Goodson K. E., "Modeling of Two-Phase Microchannel Heat Sinks for VLSI Chips", *Proceedings of IEEE MEMS Workshop*, Jan. 2001, pp. 422-426.

[5.1] Carey V., *Liquid-Vapor Phase-Change Phenomena*, Hemisphere Publishing Corporation, 1992, pp. 483-485.

[5.2] Jiang L., Wong M., Zohar Y., "Phase Change in Microchannel Heat Sink Under Forced Convection Boiling", *Proceedings of IEEE MEMS Workshop*, Jan. 2000, pp. 397-402.

[5.3] Peng X. F., Wang B. X., "Forced Convection and Flow Boiling Heat Transfer for Liquid Flowing Through Microchannels", *International Journal of Heat and Mass Transfer*, 1993, Vol. 36, No. 14, pp. 3421-3427.

[5.4] Kenning D. B. R., Yan Y., "Saturated Flow Boiling of Water in a Narrow Channel: Experimental Investigation of Local Phenomena", *IChemE Transactions A, Chemical Engineering Research and Design 79 (A4)*, May 2001, pp. 425-436.

[5.5] Peles Y. P., Yarin L. P., Hetsroni G., "Evaporating Two-Phase Flow Mechanism in Microchannels", *SPIE Symposium on Design, Test, and Microfabrication of MEMS and MOEMS*, Apr. 1999, Vol. 3680, pp. 226-229.

[6.1] Peng X. F., Hu H. Y., Wang B. X., "Boiling Nucleation During Liquid Flow in Microchannels", *International Journal of Heat and Mass Transfer*, 1998, Vol. 41, No. 1, pp. 101-106.

[6.2] Peng X. F., Tien Y., Lee D. J., "Bubble Nucleation in Microchannels: Statistical Mechanics Approach", *International Journal of Heat and Mass Transfer*, Aug. 2001, Vol. 44, No. 15, pp. 2957-2964.

[6.3] Blander M., Katz J. L., "Bubble Nucleation in Liquids", *AIChe Journal*, Sep. 1975, Vol. 21, No. 5, pp. 833-848.

[6.4] Carey V., *Liquid-Vapor Phase-Change Phenomena*, Hemisphere Publishing Corporation, 1992, pp. 169-211.

[6.5] Hsu Y. Y., "On the Size Range of Active Nucleation Cavities on a Heating Surface", *ASME Journal of Heat Transfer*, Aug. 1962, pp. 207-216.

[6.6] Thome J. R., *Enhanced Boiling Heat Transfer*, Hemisphere Publishing Corporation, 1989.

[6.7] Bergles A. E., "Enhanced Heat Transfer Techniques for High-Heat-Flux Boiling", *Proceedings of SPIE High Heat Flux Engineering*, 1992, Vol. 1739, pp. 2-16.

[6.8] Liter S. G., Kaviany M., "CHF Enhancement by Modulated Porous-Layer Coating", *Proceedings of the ASME IMECE*, Nov. 1998, HTD Vol. 361-1, pp. 165-173.

[6.9] Ramaswamy C., Johsi Y., Nakayama W., Johnson W. B., "High-Speed Visualization of Boiling from an Enhanced Structure", *International Journal of Heat and Mass Transfer*, Nov. 2002, Vol. 45, No. 24, pp. 4761-4771.

[6.10] Klausner J. F., Mei R., "Suppression of Nucleation Sites in Flow Boiling", *Proceedings of Convective Flow Boiling*, 1995, pp. 155-160.

[6.11] Lin L., Pisano A. P., "Bubble Formation on a Micro Line Heater", *Proceedings of ASME Micromechanical Sensors, Actuators, and Systems*, 1991, DSC-Vol. 32, pp. 147-163.

[6.12] Oh S.-D., Seung S. S., Kwak H. Y., "A Model of Bubble Nucleation on a Micro Line Heater", *Journal of Heat Transfer*, Feb. 1999, Vol. 121, pp. 220-225.

[6.13] Kwak H.-Y., Panton R L., "Gas Bubble Formation in Nonequilibrium Water-Gas Solutions", *Journal of Chemical Physics*, May 1983, Vol. 78, No. 9, pp. 5795-5799.

[6.14] Kwak H.-Y., Lee S., "Homogeneous Bubble Nucleation Predicted by a Molecular Interaction Model", *ASME Journal of Heat Transfer*, Aug. 1991, Vol. 113, pp. 714-721.

[6.15] Braff R. A., Gerhardt A. L., Schmidt M. A., Gray M. L., Toner M., "A Microbubble-Power Boparticle Actuator", *Proceedings of Solid-State Sensor, Actuator, and Microsystems Workshop*, Jun. 2002, pp. 138-141.

Index

Druck: betz-druck GmbH, D-64291 Darmstadt
Verarbeitung: Buchbinderei Schäffer, D-67269 Grünstadt